Memoirs of the American Mathematical Society

Number 355

Gregory T. Adams

The bilateral Bergman shift

Published by the

AMERICAN MATHEMATICAL SOCIETY

Providence, Rhode Island, USA

September 1986 · Volume 63 · Number 355 (end of volume)

MEMOIRS of the American Mathematical Society

SUBMISSION. This journal is designed particularly for long research papers (and groups of cognate papers) in pure and applied mathematics. The papers, in general, are longer than those in the TRANSACTIONS of the American Mathematical Society, with which it shares an editorial committee. Mathematical papers intended for publication in the Memoirs should be addressed to one of the editors:

Ordinary differential equations, partial differential equations, and applied mathematics to JOEL A. SMOLLER, Department of Mathematics, University of Michigan, Ann Arbor, MI 48109

Complex and harmonic analysis to LINDA PREISS ROTHSCHILD, Department of Mathematics, University of California at San Diego, La Jolla, CA 92093

Abstract analysis to VAUGHAN F. R. JONES, Department of Mathematics, University of California, Berkeley, CA 94720

Classical analysis to PETER W. JONES, Department of Mathematics, Box 2155 Yale Station, Yale University, New Haven, CT 06520

Algebra, algebraic geometry, and number theory to LANCE W. SMALL, Department of Mathematics, University of California at San Diego, La Jolla, CA 92093

Geometric topology and general topology to ROBERT D. EDWARDS, Department of Mathematics, University of California, Los Angeles, CA 90024

Algebraic topology and differential topology to RALPH COHEN, Department of Mathematics, Stanford University, Stanford, CA 94305

Global analysis and differential geometry to TILLA KLOTZ MILNOR, Department of Mathematics, Hill Center, Rutgers University, New Brunswick, NJ 08903

Probability and statistics to RONALD K. GETOOR, Department of Mathematics, University of California at San Diego, La Jolla, CA 92093

Combinatorics and number theory to RONALD L. GRAHAM, Mathematical Sciences Research Center, AT&T Bell Laboratories, 600 Mountain Avenue, Murray Hill, NJ 07974

Logic, set theory, and general topology to KENNETH KUNEN, Department of Mathematics, University of Wisconsin, Madison, WI 53706

All other communications to the editors should be addressed to the Managing Editor, WILLIAM B. JOHNSON, Department of Mathematics, Texas A&M University, College Station, TX 77843-3368

PREPARATION OF COPY. Memoirs are printed by photo-offset from camera-ready copy prepared by the authors. Prospective authors are encouraged to request a booklet giving detailed instructions regarding reproduction copy. Write to Editorial Office, American Mathematical Society, Box 6248, Providence, RI 02940. For general instructions, see last page of Memoir.

SUBSCRIPTION INFORMATION. The 1986 subscription begins with Number 339 and consists of six mailings, each containing one or more numbers. Subscription prices for 1986 are $214 list, $171 institutional member. A late charge of 10% of the subscription price will be imposed on orders received from nonmembers after January 1 of the subscription year. Subscribers outside the United States and India must pay a postage surcharge of $18; subscribers in India must pay a postage surcharge of $15. Each number may be ordered separately; *please specify number* when ordering an individual number. For prices and titles of recently released numbers, see the New Publications sections of the NOTICES of the American Mathematical Society.

BACK NUMBER INFORMATION. For back issues see the AMS Catalogue of Publications.

Subscriptions and orders for publications of the American Mathematical Society should be addressed to American Mathematical Society, Box 1571, Annex Station, Providence, RI 02901-1571. *All orders must be accompanied by payment.* Other correspondence should be addressed to Box 6248, Providence, RI 02940.

MEMOIRS of the American Mathematical Society (ISSN 0065-9266) is published bimonthly (each volume consisting usually of more than one number) by the American Mathematical Society at 201 Charles Street, Providence, Rhode Island 02904. Second Class postage paid at Providence, Rhode Island 02940. Postmaster: Send address changes to Memoirs of the American Mathematical Society, American Mathematical Society, Box 6248, Providence, RI 02940.

Printed in the United States of America.

The paper used in this journal is acid-free and falls within the guidelines established to ensure permanence and durability.

Table of Contents

ABSTRACT

The analogy between the classical Toeplitz operators on the circle and the Bergman Toeplitz operators on the disc is extended to an analogy between the classical multiplication operators M_φ on the circle and a new family of operators C_φ on $L^2_h(D)$ where $L^2_h(D)$ is the collection of square integrable complex valued harmonic functions on the disc D (equipped with area measure). It is shown that in some sense the difference between M_φ and C_φ is Hilbert-Schmidt exactly when the analytic and coanalytic parts of φ are in Dirichlet space.

The commutant of M_z, the classical bilateral shift is identical with the algebra of bounded functions on the circle. The commutant of C_z, the bilateral Bergman shift is also identified with an algebra of bounded functions on the circle with multiplication in the algebra corresponding to pointwise multiplication of functions on the circle. This algebra has the form $H^\infty(T) \oplus \bar{B}$ where B denotes the Zygmund class and the bar denotes complex conjugation.

The algebra is studied and the maximal ideal space is shown to correspond to the corona of H^∞. It is also shown that the bilateral Bergman shift has no square root.

1980 Mathematics Subject Classification 47B37

Library of Congress Cataloging-in-Publication Data

Adams, Gregory T.

The bilateral Bergman shift.

(Memoirs of the American Mathematical Society, ISSN 0065-9266; no. 355)

Bibliography: p.

1. Shift operators (Operator theory) 2. Toeplitz operators. I. Title. II. Series.

QA3.A57 no. 355 [QA329.2] 510 s [515.7'246] 86-17404

ISBN 0-8218-2417-1

§1. Introduction

For several years quite a deal of attention has been paid to the study of Bergman Toeplitz operators. If ϕ is a bounded function on the unit disc, then the corresponding Toeplitz operator on the Bergman space will be denoted by B_ϕ. These operators arose as a natural generalization of the classical Toeplitz operators $T_{\tilde{\phi}}$, acting on the Hardy space H^2. In the course of sudying the Bergman Toeplitz operators I was led to consider another type of operator. It is the purpose of this introduction to give the motivation for looking at these new operators. Relevant definitions will be postponed till later.

First let's recall the way in which the operators B_ϕ generalize the operators $T_{\tilde{\phi}}$. If $\tilde{\phi}$ is a bounded measurable function on the unit circle, then $T_{\tilde{\phi}}$ is the compression of multiplication by $\tilde{\phi}$ to H^2. If $\phi: D \to \mathbb{C}$ is a bounded measurable function, then B_ϕ is the compression of multiplication by ϕ to $L_a^2(D)$, the Bergman space. The functions $\tilde{\phi}$ and ϕ are called the symbols of $T_{\tilde{\phi}}$ and B_ϕ, respectively. If ϕ is continuous on the closed disc with boundary values $\tilde{\phi}$, then B_ϕ is unitarily equivalent to an operator on H^2 which differs from $T_{\tilde{\phi}}$ by a compact operator.

However, there are many differences between the two kinds of Toeplitz operators. The matrix of $T_{\tilde{\phi}}$ with respect to the standard orthonormal basis of H^2 may be defined in terms of the Fourier coefficients $\{\hat{\tilde{\phi}}(n): n \in \mathbb{Z}\}$ of $\tilde{\phi}$. If a sequence, $\{a_n : n \in \mathbb{Z}\}$ of complex numbers is used to define a formal Toeplitz operator T, then T will be bounded if and only if there is a bounded function $\phi: \mathbb{T} \to \mathbb{C}$ such that $a_n = \hat{\tilde{\phi}}(n)$ for $\hat{\phi}(n)$

Received by the editors December 10, 1985.
* Partially supported by the Alexander von Humboldt-Stiftung

for each n. In this case $T = T_{\tilde{\phi}}$. For Bergman Toeplitz operators the situation is different. One can define in a perfectly reasonable way bounded operators B_ϕ for certain unbounded functions $\phi: D \to \mathbb{C}$. One example occurs by taking $\phi(z) = 1/z$. The only compact classical Toeplitz operator is the zero operator. For the Bergman space B_ϕ is compact whenever ϕ is continuous and vanishes on T, the boundary of D. If one restricts attention to those operators B_ϕ whose symbol ϕ is a harmonic function on the disc D, then the differences above disappear. There are two other good reasons for considering only harmonic symbols. Firstly, the Poisson kernel gives an identification of bounded functions on the circle with bounded harmonic functions on the disc. This is the function theoretic reason. The operator theoretic reason is the following: The weak closure of the power of T_z and its adjoint T_z^* are the classical Toeplitz operators. Similarly, the weak closure of the powers of B_z and its adjoint B_z^* is the class of Bergman Toeplitz operators with harmonic symbols.

For analytic symbols ϕ, B_ϕ is just multiplication by ϕ, and Axler [1] has determined spectral properties for such operators on more general domains than the disc. If ϕ is real valued and harmonic, then McDonald and Sundberg [4] have determined the spectrum of B_ϕ. In each case the results agree with the classical case $T_{\tilde{\phi}}$, where $\tilde{\phi}$ is the boundary value function of ϕ. Except for quite specific examples, these are the only symbols ϕ for which the spectrum of B_ϕ is known.

In the classical case let $\phi: T \to \mathbb{C}$ be a bounded function. $T_{\tilde{\phi}}$ may be thought of as operating on the Hardy space by a two-step process. H^2 is a Hilbert space subspace of $L^2(T)$ and hence there is an orthonormal projection $P: L^2(T) \to H^2$. If $f \in H^2$ and $\tilde{\phi}: T \to \mathbb{C}$ is bounded, then the pointwise product $(\tilde{\phi}f)(e^{i\theta}) = \tilde{\phi}(e^{i\theta})f(e^{i\theta})$ makes sense as an element of $L^2(T)$. The corresponding operator is denoted by $M_{\tilde{\phi}}$. $T_{\tilde{\phi}}$ is then defined by $T_{\tilde{\phi}}f = P(M_{\tilde{\phi}}f)$. Suppose now that $\tilde{\phi}$ is continuous and $0 \notin \tilde{\phi}(T)$.

It can be shown that $T_{\tilde{\phi}}$ is a Fredholm operator with index $i(T_{\tilde{\phi}})$ equal to minus the winding number of $\tilde{\phi}$ about 0. If the winding number is non-zero, then $T_{\tilde{\phi}}$ is not invertible.

If $\phi: D \to \mathbb{C}$ has continuous boundary values, then all which has been said above applies to B_ϕ. In the classical case it turns out that $T_{\tilde{\phi}}$ has always either a trivial kernel or a trivial cokernel. Thus $T_{\tilde{\phi}}$ is invertible for $\tilde{\phi}: T \to \mathbb{C}$ continuous, whenever $0 \notin \tilde{\phi}(T)$ and the winding number of $\tilde{\phi}$ about 0 is zero. This is a result of Coburn [3]. The key ingredients of the proof involve exploiting the way that the space of functions H^2 sits as a subspace in $L^2(T)$, and exploiting the two-step definition of $T_{\tilde{\phi}}$. It would seen as if the natural analog to $L^2(T)$ in the Bergman space case would be $L_h^2(D)$, the space of square integrable harmonic functions. $L_a^2(D)$ is related to $L_h^2(D)$ in much the same way H^2 is related to $L^2(T)$. In place of the multplication operator $M_{\tilde{\phi}}$, one should consider the compression C_ϕ of multplication by ϕ to $L_h^2(D)$.

The class of operators $C_\phi: L_h^2(D) \to L_h^2(D)$ is a very natural generalization of the multiplication operators $M_\phi: L^2(T) \to L^2(T)$. These operators have independent interest in their own right but also seem to be important to the study of Bergman Toeplitz operators.

The operator C_z is the Bergman bilateral shift. Its properties will be studied here in detail. Indeed, the most interesting result of this work is a concrete realization of the commutant of $C_{\bar{z}}$ as an algebra of functions on the unit circle.

In Section Two the Bergman Toeplitz operators are defined and contrasted with the classical Toeplitz operators. It is shown that there is a Schur multiplier which multiplies the class of Toeplitz operators isometrically onto the class of Bergman Toeplitz operators with harmonic symbols. Further, if $V: H^2(T) \to L_a^2(D)$ is the "natural" isomorphism, and $\phi: D \to \mathbb{C}$ is harmonic and bounded with boundary values also denoted by ϕ, then

$V^*B_\phi V - T_\phi$ is Hilbert-Schmidt precisely when the analytic and coanalytic parts of ϕ are in Dirichlet class on the disc D.

In Section Three the Bergman bilateral shift C_z and the operators C_ϕ are introduced. The operators C_ϕ are compared to the multiplication operators on $L^2(T)$ and theorems analogous to those in Section Two are proven.

In Section Four a certain family of Besov spaces is introduced. Some equivalent formulations of these spaces are given. These spaces are a main ingredient of the algebras defined in Section Five.

Section Five contains a description of the commutant of an arbitrary injective bilateral shift. This description is used to describe the commutant of C_z as an algebra of functions on the unit circle T.

In Section Six the algebra of the commutant is investigated. The maximal ideal space is determined and the spectrum of certain elements of the algebra is investigated. Additionally it is shown that C_z has no square root.

In Section Seven some final comments are made.

I would like to thank John Conway for his support during the development of these ideas. Talks with G.Bennett also proved very useful. Finally I acknowledge the support of the Alexander von Humboldt-Stiftung for its kind financial aid.

§2. THE BERGMAN TOEPLITZ OPERATORS

The Bergman Toeplitz operators B_ϕ have been investigated for some time as a natural family of operators to study after considering the classical Toeplitz operators $T_{\tilde{\phi}}$. If D is the open unit disc in the complex plane $\mathbb{C}$, equipped with normalized area measure dA, then the Bergman space $L_a^2(D)$ is the collection of square integrable analytic functions on D. Let $L^2(D)$ denote the space of square integrable functions on D, again equipped with area measure. Both $L^2(D)$ and $L_a^2(D)$ are Hilbert spaces with respect to the obvious norm (see Conway [2], p. 176). Let $P: L^2(D) \to L_a^2(D)$ denote the orthogonal projection. If $\phi \in L^\infty(D)$, the space of bounded measurable functions on D, then define $B_\phi: L_a^2(D) \to L_a^2(D)$ by putting $B_\phi f = P(\phi f)$ for each $f \in L_a^2(D)$. B_ϕ ist the Bergman Toeplitz operator with symbol ϕ and is seen to be the compression of multplication by ϕ to the Bergman space. The Bergman Toeplitz operators share many properties with the classical Toeplitz operators. In this section some of the similarities and differences will be discussed.

If $H^2 = H^2(T)$ denotes the space of square integrable functions on the unit circle T, whose negative Fourier coefficients vanish, then H^2 is a subspace of $L^2(T)$, the space of square integrable functions on T. In both cases the circle is equipped with normalized arc length measure. The nth Fourier coefficient of an element $f \in L^2(T)$ is of course just the quantity $f(n) = \frac{1}{2\pi}\int_0^{2\pi} f(e^{i\theta})e^{-in\theta}d\theta$. If $\tilde{\phi} \in L^\infty(T)$, the space of bounded measurable functions on T, then the Toeplitz operator $T_{\tilde{\phi}}: H^2 \to H^2$ is defined by the equation $T_{\tilde{\phi}}f = P_H\tilde{\phi}f$, where $P_H: L^2(T) \to H^2(T)$ is the orthogonal projection. For a discussion of these operators see Douglas [3], Chapter 7.

The operators B_ϕ share the usual algebraic properties of Toeplitz operators. Thus if $\gamma, \phi, \phi_1, \phi_2 \in L^\infty(D)$ and if γ happen also to be analytic, then

$$B_{\phi_1+\phi_2} = B_{\phi_1} + B_{\phi_2} \; ,$$

$$B_{\phi\gamma} = B_\phi \, B_\gamma \; ,$$

$$B_{\phi\bar{\gamma}} = B_{\bar{\gamma}} \, B_\phi \; ,$$

$$B^*_\phi = B_{\bar{\phi}} \; .$$

Here "*" denotes adjoint and the bar "-" denotes complex conjugation.

There is a natural way of comparing the Bergman Toeplitz operators with the classical Toeplitz operators. $H^2(T)$ has an orthonormal basis consisting of the functions $1, e^{i\theta}, e^{2i\theta}, \ldots$. The corresponding orthonormal basis for $L^2_a(D)$ consists of the functions $1, \sqrt{2}z, \sqrt{3}z^2, \ldots$. Here $e^{i\theta}$ and z are the independent variables on T and D, respectively. Let $V\colon H^2 \to L^2_a(D)$ be the unitary map defines by $V(e^{in\theta}) = \sqrt{n+1}\, z^n$. The operator V provides a way of quantitatively comparing the two classes of Toeplitz operators. Thus an operator B_ϕ will be compared with a classical Toeplitz operator via the unitarily equivalent operator $V^*B_\phi V$, which lives on $H^2(T)$. More will be said about this later.

Letting z denote the independent variable on D, the operator B_z is of particular interest. B_z is generally known as the Bergman shift. The Bergman shift is a natural generalization of the unilateral shift $T_{e^{i\theta}}$, which is just multiplication by the independent variable on $H^2(T)$. B_z is subnormal and has the closed unit disc as its spectrum, with the unit circle as its essential spectrum. As shall be seen later in this section, the difference $T_{e^{i\theta}} - V^*B_zV$ is Hilbert-Schmidt and hence compact, but this difference is not trace class. B_z is of course just multiplication by the bounded analytic function z. Axler [1] has determined the relevant spectral properties for analytic multiplication operators on Bergman spaces of more general domains in $\mathbb{C}$. For harmonic symbols ϕ, the question of determining even the spectrum of B_ϕ has not yet been answered.

Let $L_h^\infty(D)$ denote the space of bounded harmonic functions on D. The Poisson kernel induces an isometry $L^\infty(T) \to L_h^\infty(D)$. Since each function in $L_h^\infty(T)$ possesses boundary values almost everywhere on T, and since the functions in $L_h^\infty(D)$ obey the maximum principle, it follows that this map is also surjective. A function in $L_h^\infty(D)$ will be denoted by ϕ and its boundary values by $\tilde{\phi}$.

There are good reasons for restricting the study of operators B_ϕ to those operators whose symbols ϕ are in $L_h^\infty(D)$. The class $\{T_{\tilde{\phi}}: \tilde{\phi} \in L^\infty(\phi)\}$ is the weakly closed span of the powers $\{T^n_{e^{i\theta}}, T^{*n}_{e^{i\theta}}: n \in \mathbb{Z}_+\}$, of the unilateral shift $T_{e^{i\theta}}$ and its adjoint $T^*_{e^{i\theta}}$. In the same way $\{B_\phi: \phi \in L_h^\infty(D)\}$ is the weakly closed span of the positive powers of the Bergman shift B_z and its adjoint. The proof of this will be postponed to Proposition 2.3. The only compact classical Toeplitz operator $T_{\tilde{\phi}}$, is the zero operator with $\tilde{\phi} = 0$. For a general Bergman Toeplitz operator B_ϕ, this is not true. In fact, B_ϕ ist compact whenever φ vanishes in some sense in a neighborhood of T. If $\phi \in L_h^\infty(D)$, however, then B_ϕ is compact only when $\phi \equiv 0$. In the interest of the analogy between the Bergman Toeplitz operators and the classical Toeplitz operators, it would seem natural to restrict the class of symbols for Bergman Toeplitz operators to the space of bounded harmonic functions on the disc. From now on, all symbols ϕ for Bergman Toeplitz operators will be taken from $L_h^\infty(D)$. In light of the identification $L^\infty(T) \cong L_h^\infty(D)$ via the Poisson kernel, there will be no ambiguity if the symbol ϕ is used to denote both an element of $L_h^\infty(D)$ and the element of $L^\infty(T)$ which represents its boundary values.

The concept of Schur multiplication is useful in relating the family $\{B_\phi: \phi \in L_h^\infty(D)\}$ to the family $\{T_\phi: \phi \in L^\infty(T)\}$. If F is a family of operators on a Hilbert space $\mathcal{H}$ with basis $\{e_0, e_1, e_2, \ldots\}$, then each F in $\mathcal{F}$ has a matrix representation $(f_{jk})_{jk=0}^{\infty}$ with respect to this basis given by $f_{jk} = \langle Fe_k, e_j \rangle$, where $\langle , \rangle$ is the inner product on $\mathcal{H}$. A Schur multiplier for the familiy $\mathcal{F}$ is a matrix of coefficients (m_{ij}) such that for each representative of an element of $\mathcal{F}$, the matrix (a_{jk}) defined by $a_{jk} = m_{jk} f_{jk}$ corresponds to the matrix of a bounded operator on $\mathcal{H}$.

The following proposition asserts that there is a Schur multiplier which carries the collection $\{T_\phi : \phi \in L^\infty(T)\}$ onto $\{B_\phi : \phi \in L_h^\infty(D)\}$ with the correspondence mapping T_ϕ to B_ϕ.

<u>2.1 Proposition</u> If $\phi \in L_h^\infty(D)$ and if (b_{jk}) is the matrix of B_ϕ with respect to the standard basis of $L_a^2(D)$, and if (a_{jk}) is the matrix of T_ϕ with respect to the standard basis of $H^2(T)$, then these matrices are related by the formula

$$b_{jk} = \begin{cases} \sqrt{\dfrac{j+1}{k+1}}\, a_{jk} & \text{if } j \leq k \\ \sqrt{\dfrac{k+1}{j+1}}\, a_{jk} & \text{if } k < j . \end{cases}$$

<u>Proof</u>. If ϕ $L_h^\infty(D)$, then ϕ has the expansion

$$\phi(z) = \sum_{n<0} a_n \bar{z}^{|n|} + \sum_{n \geq 0} a_n z^n ,$$

where convergence is uniform on compact subsets of D. The boundary value function has the Fourier expansion

$$\phi(e^{i\theta}) = \sum_{n=-\infty}^{\infty} a_n e^{in\theta} .$$

If (b_{jk}) is the matrix of B_ϕ with respect to the standard basis for $L_a^2(D)$, and if $\langle , \rangle$ is the inner product on $L^2(D)$, then

$$\begin{aligned} b_{jk} &= \langle B_\phi z^k \sqrt{k+1},\ z^j \sqrt{j+1}\rangle \\ &= \langle \phi(z) z^k \sqrt{k+1},\ z^j \sqrt{j+1}\rangle \\ &= \frac{1}{\pi}\int \phi(z)\ z^k \sqrt{k+1}\ \bar{z}^j \sqrt{j+1}\ dA(z) \\ &= \frac{c_{jk}}{\pi}\int_0^1\int_0^{2\pi} \phi(re^{i\theta}) e^{i(k-j)\theta}\ r^{j+k+1}\ d\theta\ dr \ , \end{aligned}$$

where $c_{jk} = \sqrt{k+1}\sqrt{j+1}$. Substitute now the series expansion for $\phi(z)$ with $z = re^{i\theta}$. For each $r < 1$ the series converges uniformly and absolutely in θ so that orders of summation and integration may be exchanged. Hence

$$b_{jk} = \frac{c_{jk}}{\pi}\int_0^1 \sum_{n\in\mathbb{Z}} \int_0^{2\pi} a_n\ r^{|n|}\ e^{ir\theta}\ e^{i(k-j)\theta}\ d\theta\ r^{j+1+1}\ dr \ .$$

For fixed r, the integral in θ is 0 if $n \neq j-k$, so

$$\begin{aligned} b_{jk} &= 2c_{jk}\int_0^1 a_{j-k}\ r^{|j-k|+j+k+1}\ dr \\ &= \frac{2c_{jk}}{|j-k| + j + k + 2}\ a_{j-k} \ . \end{aligned}$$

Now $|j-k| + j + k + 2 = 2(\max\{j,k\} + 1)$. Hence

$$\begin{aligned} b_{jk} &= \frac{\sqrt{j+1}\ \sqrt{k+1}}{1 + \max\{j,k\}}\ a_{j-k} \\ &= \begin{cases} \sqrt{\dfrac{j+1}{k+1}}\ a_{j-k} & \text{if } j \leq k \ , \\ \sqrt{\dfrac{k+1}{j+1}}\ a_{j-k} & \text{if } j > k \ . \end{cases} \end{aligned}$$

A similar but simpler computation shows that the matrix (a_{jk}) of T_ϕ is given by $a_{jk} = a_{j-k}$ which proves the proposition. □

The coefficient matrix

$$m_{jk} = \begin{cases} \sqrt{\frac{j+1}{k+1}} & \text{if } j \leq k , \\ \sqrt{\frac{k+1}{j+1}} & \text{if } j > k \end{cases}$$

will so often be used that from now on the symbol m_{jk} will bereserved to denote it.

If $V: H^2(T) \to L_a^2(D)$ denotes the previously defined isomorphism, and if $\phi \in L_h^\infty(D)$ has the expansion $\phi(z) = \sum_{n<0} a_n \bar{z}^{|n|} + \sum_{n\geq 0} a_n z^n$, then the previous proof shows that the matrix (c_{jk}), of $T_\phi - V^*B_\phi V$ with respect to the standard basis of $H^2(T)$ is given by

$$c_{jk} = (1 - m_{jk})a_{j-k} .$$

A classical Toeplitz operator has a matrix (a_{jk}), where $a_{jk} = a_{j-k}$ for some sequence $\{a_n\}_{n\in \mathbb{Z}}$. A matrix of this form corresponds to a bounded operator if and only if $\sum a_n e^{in\theta}$ is the Fourier expansion of a bounded function ϕ on T. In this case the matrix corresponds to the Toeplitz operator T_ϕ. The previous proposition demonstrates that if $\phi \in L_h^\infty(D)$, then the matrix for B_ϕ has the form of the Schur multiplier (m_{jk}) times the matrix of T_ϕ. The following proposition shows that any matrix of a bounded operator which has the formal form of a Bergman Toeplitz operator, corresponds to an operator B_ϕ where $\phi \in L_h^\infty(D)$.

Before stating the proposition some notation is in order. An $n \times n$ matrix may be viewed as an operator on the Hilbert space $\mathbb{C}^n$ in the obvious way. The operator norm of such a matrix will be denoted by $\|\ \|$. The norm of an infinite dimensional matrix, viewed as an operator on an infinite dimensional Hilbert space, is just the supremum of its finite dimensional submatrices. If this supremum is infinite, then the matrix in questions does not correspond to the matrix of a bounded operator. In fact, if $(c_{jk})_{jk=0}^\infty$

is an infinite dimensional matrix, then

$$\|(c_{jk})_{jk=0}^{\infty}\| = \sup_n \|(c_{jk})_{jk=0}^{n}\| .$$

The proof of this last fact is elementary and will be omitted.

2.2 Proposition If $B = (b_{jk})_{jk=0}^{\infty}$ is a matrix of the form $b_{jk} = m_{jk}\, a_{j-k}$ for some sequence $\{a_n\}_{n\in\mathbb{Z}}$, and if B corresponds to the matrix of a bounded operator also denoted B, then

$$\phi(z) = \sum_{n<0} a_n \bar{z}^{|n|} + \sum_{n\geq 0} a_n z^n$$

defines a bounded harmonic function on D and $B \cong B_\phi$.

Proof. In light of Proposition 2.1, it suffices to show that $\sum_n a_n e^{in\theta}$ is the Fourier expansion of a bounded function on T. Equivalently it suffices to show that the matrix $A = (a_{jk})_{jk=0}^{\infty}$ defined by $a_{jk} = a_{j-k}$, corresponds to a bounded operator. Let A_n be the $n\times n$ matrix $(a_{jk})_{jk=0}^{n-1}$. A corresponds to a bounded operator exactly when $\sup_n \|A_n\| < \infty$.

Let n be fixed and consider the $n\times n$ submatrix B_ℓ of B defined by $B_\ell = (c_{jk})_{jk=0}^{n-1}$, where $c_{jk} = b_{j+\ell,k+\ell} = m_{j+\ell,k+\ell}\, a_{jk}$. On the one hand, $\|B_\ell\| \leq \|B\|$ for each positive integer ℓ. On the other hand, $\lim_{\ell\to\infty} m_{j+\ell,k+\ell} = 1$ for each pair of positive integers (j,k). Therefore the coordinate-wise limit of the matrices B_ℓ as $\ell\to\infty$ is A_n. Therefore $\|A_n\| \leq \sup_\ell \|B_\ell\| \leq \|B\|$. The norms $\|A_n\|$ are uniformly bounded. Therefore $\|A\| < \infty$. □

The family $\{T_\phi : \phi\in L^\infty(T)\}$ is the weakly closed span of the positive powers of the unilateral shift $T_{e^{i\theta}}$ and its adjoint. The following proposition lends weight to the analogy between the families $\{T_\phi : \phi \in L^\infty(T)\}$ and $\{B_\phi : \phi \in L^\infty(T)\}$.

<u>2.3 Proposition</u> $\{B_\phi : \phi \ L_h^\infty(D)\}$ is the weak operator topology closure of the positive powers of the Bergman shift B_z and its adjoint $B_{\bar{z}}$.

<u>Proof</u>. Let $F = \{B_\phi : \phi \in L_h^\infty(D)\}$ and let $G = \{B_\phi : \phi$ is a polynomial in z and $\bar{z}\}$. Here a polynomial in z and $\bar{z}$ means a linear combination of powers of z and $\bar{z}$. In light of the algebraic properties of Bergman Toeplitz operators G is the linear span of the positive powers of B_z and $B_{\bar{z}}$. It will suffice to show that in the weak operator topology F is closed and G is dense in F.

Suppose $B = \lim B_{\phi_\ell}$ WOT, where $\phi_\ell \in L_h^\infty(D)$. Let $(b_{jk\ell})_{jk=0}^\infty$ be the matrix representation of B_{ϕ_ℓ}. If (b_{jk}) is the matrix of B, then the WOT convergence implies $\lim_\ell b_{jk\ell} = b_{jk}$. By Proposition 2.1 there are sequences $\{a_n\}_{n\in\mathbb{Z}}$ for each ℓ such that $b_{jk\ell} = m_{jk}\, a_{(j-k),\ell}$. This shows that there is a single sequence $\{a_n\}$ such that $a_n = \lim_\ell a_{n\ell}$ and such that $b_{jk} = m_{jk}\, a_{j-k}$ for each pair of positive integers (j,k). By Proposition 2.2 there is a $\phi \in L_h^\infty(D)$ such that $B = B_\phi$. This shows that F is WOT closed.

Now assume $\phi \in L_h^\infty(D)$. For $r \in (0,1)$ define $\phi_r : D \to \mathbb{C}$ by $\phi_r(z) = \phi(rz)$. Each function ϕ_r is harmonic, and by the maximum principle $\|\phi_r\|_\infty \leq \|\phi\|_\infty$. Moreover on any compact set $K \subset D$, ϕ_r converges uniformly to ϕ as $r \to 1$. Since ϕ_r has a uniformly absolutely convergent expansion in powers of z and $\bar{z}$ it follows that B_{ϕ_r} is in the WOT closure of G. To show that G is dense in F, it suffices to show that

$$B_\phi = \lim_{r\to 1^-} B_{\phi_r} \quad \text{WOT} .$$

Let $f,g \in L_a^2(D)$ and fix $\varepsilon > 0$. If $t \in (0,1)$, let $D_t = \{\lambda \in \mathbb{C}: |\lambda| \leq t\}$ and let $E_t = D \setminus D_t$. Choose $t_0 < 1$ so that

$$\left(\int_{E_t} |f|^2 \, dA\right)^{1/2} < \frac{\varepsilon}{4\|\phi\|_\infty}$$

and

$$\left(\int_{E_t} |g|^2 \, dA\right)^{1/2} < 1 .$$

Next choose $r_0 < 1$ so that if $r > r_0$, then

$$\sup_{z \in D_t} |\phi_r(z) - \phi(z)| < \frac{\varepsilon}{2\,\|f\|_2\,\|g\|_2} .$$

Therefore if $r > r_0$, then

$$\begin{aligned}
|\langle B_\phi f,g\rangle - \langle B_{\phi_r} f,g\rangle| &= |\langle B_{\phi-\phi_r} f,g\rangle| \\
&= \left|\int_D (\phi-\phi_r) f\bar{g} \, dA\right| \\
&\leq \left|\int_{D_t} (\phi-\phi_r) f\bar{g} \, dA\right| + \left|\int_{E_t} (\phi-\phi_r) f\bar{g} \, dA\right| \\
&\leq \sup_{D_t} |\phi(z)-\phi_r(z)| \, \|f\|_2 \, \|g\|_2 \\
&\qquad + \sup_{E_t} |\phi(z)-\phi_r(z)| \left(\int_{E_t} |f|^2 \, dA\right)^{1/2} \\
&\qquad \times \left(\int_{E_t} |g|^2 \, dA\right)^{1/2} \\
&\leq \frac{\varepsilon}{2\,\|f\|_2\,\|g\|_2} \cdot \|f\|_2 \, \|g\|_2 + 2\,\|\phi\|_\infty \frac{\varepsilon}{4\|\phi\|_\infty} 1 \\
&= \varepsilon .
\end{aligned}$$

Clearly $\lim_{r\to 1^-} \langle B_{\phi_r} f,g\rangle = \langle B_\phi f,g\rangle$ for each $f,g \in L_a^2(D)$. Therefore $B_\phi = \lim_{r\to 1^-} B_{\phi_r}$ WOT. □

If $V: H^2 \to L_a^2(D)$ is the previously defined isomorphism, let $D_z : H^2(T) \to H^2(T)$ be the difference $T_{e^{i\theta}} - V^*B_zV$. The matrix of D_z with respect to the standard basis has the form $(d_{jk})_{jk=0}^{\infty}$, where

$$d_{jk} = \begin{cases} 0 & \text{if } j-k \neq 1 , \\ 1 - \dfrac{j}{j+1} & \text{if } j-k = 1 . \end{cases}$$

Thus (d_{jk}) is zero on all but the submain diagonal and there the entries tend towards 0. In other words, D_z is compact. Some algebra shows that $(T_{e^{i\theta}})^n - V^*B_z^nV$ is compact for each positive integer n. Applying adjoints gives the same result for $T^n_{e^{-i\theta}} - V^*B_{\bar{z}}^nV$. This shows that $T_\phi - V^*B_\phi V$ is compact for any polanomial ϕ in z and $\bar{z}$. Finally, taking norm limits, one obtains the following result.

<u>2.4 Proposition</u> If $\phi: D \to \mathbb{C}$ is a harmonic function which is continuously extendable to $\bar{D}$, then

$$D_\phi = T_\phi - V^* B_\phi V$$

is compact. □

The preceding proposition gives a condition under which $T_\phi - V^*B_\phi B$ is compact. One would prefer a more quantitative estimate for the "compactness" of this difference. If K is a compact operator on a Hilbert space H, let $|K| = (K^*K)^{1/2}$ be the absolute value of K. $|K|$ has countably many eigenvalues, $\{\lambda_j\}$, each nonnegative, and these can be arranged in a decreasing sequence $\lambda_1 \geqq \lambda_2 \geqq \ldots$ with each eigenvalue repeated as often as its multiplicity. Since K is compact, the eigenvalues λ_j tend towards 0. K is said to be trace class if the trace norm $\|K\|_1 = \sum \lambda_i$ is finite. The trace class operators form an ideal in $B(H)$, the algebra of bounded linear operators on H. Another ideal of compact operators is the Hilbert-Schmidt class. Again let K be a compact operator on a Hilbert space H and let $\{e_j\}$ be an orthonormal basis for H. K is Hilbert-Schmidt if the norm $\|K\|_2 = (\sum_{jk} |\langle ke_j, e_k\rangle|^2)^{1/2}$ is finite. A simple computation shows that the norm $\| \ \|_2$ is actually independent of the basis chosen. In a moment these definitions will be applied to the operator $K = T_{e^{i\theta}} - V^*B_zV$, but first a computation must be made.

Consider the function $f\colon [0,1] \to [0,1]$ defined by $f(\varepsilon) = (1-\varepsilon)^{1/2}$. The graph of f is concave down on $[0,1]$ and $f'(0) = \frac{1}{2}$. Hence if $\varepsilon \in (0,1)$, then $f(\varepsilon) < 1 - \frac{1}{2}\varepsilon$. Of course if $\alpha = (1-\varepsilon)^{1/2} \in (0,1)$, then $\alpha^2 \leqq \alpha$. Thus for $0 < \varepsilon < 1$, $1-\varepsilon < (1-\varepsilon)^{1/2} < 1 - \frac{1}{2}\varepsilon$.

<u>2.5 Example</u> $K = T_{e^{i\theta}} - V^*B_zV$ is Hilbert-Schmidt but not trace class.

<u>Proof</u>. K is the forward weighted shift on $H^2(T)$ defined by $ke^{in\theta} = (1 - \frac{n+1}{n+2})\, e^{i(n+1)\theta}$. K* is the backward weighted shift defined by

$$K^*e^{in\theta} = \begin{cases} 0 & \text{if } n = 0, \\ (1 - \sqrt{\frac{n}{n+1}})\, e^{i(n-1)\theta} & \text{if } n > 0. \end{cases}$$

Therefore $|K| = (K^*K)^{1/2}$ is the operator defined by

$$|K|e^{in\theta} = (1 - \sqrt{\frac{n+1}{n+2}})\, e^{in\theta} .$$

Applying the computation above to $\sqrt{\frac{n+1}{n+2}} = \sqrt{1 - \frac{1}{n+2}}$ gives

$$\|K\|_1 = \sum_{n=0}^{\infty} 1 - \frac{n+1}{n+2}$$

$$\geq \sum_{n=0}^{\infty} 1 - (1 - \frac{1}{2(n+2)})$$

$$- \sum_{n=0}^{\infty} \frac{1}{2(n+2)}$$

$$= \infty .$$

Therefore K is not trace class.

To show that K is Hilbert-Schmidt it suffices to show that

$$\|K\|_2^2 = \sum_{jk} |\langle Ke^{ij\theta}, e^{ik\theta}\rangle|^2 < \infty .$$

But

$$\langle Ke^{ij\theta}, e^{ik\theta}\rangle = \begin{cases} 0 & \text{if } j \neq k-1 , \\ 1 - \sqrt{\frac{j+1}{j+2}} & \text{if } j = k-1 . \end{cases}$$

By the computation, $1 - \sqrt{\frac{j+1}{j+2}} < 1 - (1 - \frac{1}{j+2}) = \frac{1}{j+2}$. Therefore

$$\|K\|_2^2 < \sum (\frac{1}{j+2})^2 < \infty . \qquad \square$$

The computation above, which was the key to working out Example 2.5, enables one to completely characterize those harmonic symbols $\phi \in L_h^{\infty}(D)$ for which $T_\phi - V^*B_\phi V$ is Hilbert-Schmidt.

An analytic function $\phi: D \to \mathbb{C}$ is said to be in Dirichlet class if the derivative of ϕ is square integrable with respect to area measure on D. Dirichlet class can be defined in terms of the power series expansions of analytic functions on the disc as well. If ϕ has the expansion

$\phi(z) = \sum_{n=0}^{\infty} a_n z^n$, then ϕ is in Dirichlet class exactly whenever $\sum_{n=1}^{\infty} |a_n|^2 n < \infty$.

If $\phi: D \to \mathbb{C}$ is a harmonic function, then ϕ can be expressed uniquely as $\phi(z) = \phi_a(z) + \overline{\phi}_{\bar{a}}(z)$, where $\phi_a, \phi_{\bar{a}}: D \to \mathbb{C}$ are analytic and $\phi_{\bar{a}}(0) = 0$. The functions ϕ_a and $\phi_{\bar{a}}$ are the analytic and coanalytic parts of ϕ.

<u>2.6 Theorem</u> If $\phi \in L_h^{\infty}(D)$, then $T_\phi - V^*B_\phi V$ is Hilbert-Schmidt if and only if ϕ_a and $\phi_{\bar{a}}$ are in Dirichlet class.

<u>Proof.</u> If $\phi \in L_h^{\infty}(D)$ has the expansion

$$\phi(z) = \sum_{n<0} a_n \bar{z}^{|n|} + \sum_{n\geq 0} a_n z^n$$

on D and if $(c_{jk})_{jk=0}^{\infty}$ is the matrix of $D_\phi = T_\phi - V^*B_\phi V$ with respect to the standard basis of $H^2(T)$, then by the remark following Proposition 2.1, $c_{jk} = (1 - m_{jk})a_{j-k}$, where

$$m_{jk} = \begin{cases} \sqrt{\dfrac{j+1}{k+1}} & \text{if } j \leq k\ , \\ \sqrt{\dfrac{k+1}{j+1}} & \text{if } j > k\ . \end{cases}$$

D_ϕ is Hilbert-Schmidt exactly whenever

$$\|D_\phi\|_2^2 = \sum |\langle D_\phi e_j, e_k\rangle|^2 < \infty\ .$$

But $\langle D_\phi e_j, e_k\rangle = c_{jk} = (1 - m_{jk})a_{j-k}$, so

$$\|D_\phi\|_2^2 = \sum_{jk>0} |a_{j-k}|^2 \, |1-m_{jk}|^2$$

$$= \sum_{j\geq k\geq 0} |a_{j-k}|^2 \, |1-m_{jk}|^2 + \sum_{k>j\geq 0} |a_{j-k}|^2 \, |1-m_{jk}|^2$$

$$= A + B .$$

By the computation, if $j \geq k$, then

$$1 - m_{jk} = 1 - \sqrt{\frac{k+1}{j+1}} \leq 1 - (1 - \frac{j-k}{j+1}) = \frac{j-k}{j+1} .$$

Thus,

$$A = \sum_{k=0}^{\infty} \sum_{j=k}^{\infty} |a_{j-k}|^2 \, |(1-m_{jk})|^2$$

$$= \sum_{n=0}^{\infty} \sum_{k=0}^{\infty} |a_n|^2 \, |1-m_{k+n,k}|^2$$

$$\leq \sum_{n=0}^{\infty} |a_n|^2 \sum_{n=0}^{\infty} (\frac{n}{n+k+1})^2$$

$$\leq \sum_{n=0}^{\infty} |a_n|^2 \int_n^{\infty} \frac{n^2}{x^2}\, dx$$

$$= \sum_{n=0}^{\infty} |a_n|^2 \, n .$$

On the other hand, if $j \geq k$, then

$$\frac{1}{2} (\frac{j-k}{j+1}) = 1 - (1 - \frac{1}{2} (\frac{j-k}{j+k})) < 1 - m_{jk} .$$

Applying this bound above gives $\frac{1}{8} \sum_{n=0}^{\infty} |a_n|^2 \, n \leq A$. A similar computation for B gives

$$\frac{1}{8} \sum_{n<0} |a_n|^2 \, |n| \leqq B \leqq \sum_{n<0} |a_n|^2 \, |n| \quad .$$

Putting these together completes the proof of the theorem. □

§3. THE OPERATORS C_ϕ

The classical Toeplitz operators T_ϕ are most naturally viewed as compressions of the multiplication operators $M_\phi : L^2(T) \to L^2(T)$. Multiplication operators have many easily determined properties, and some of these properties can be exploited to learn facts about the classical Toeplitz operators. The book [3] "Banach Algebra Techniques in Operator Theory" contains an exposition of Toeplitz operators in which one can see the interplay between the operators M_ϕ and the operators T_ϕ.

One of the reasons why one can learn much about the operator T_ϕ on $H^2(T)$ from the operator M_ϕ on $L^2(T)$ is the way $H^2(T)$ is contained in $L^2(T)$ as a subspace. $H^2(T)$ has the basis consisting of the vectors $1, e^{i\theta}, e^{2i\theta}, \ldots$. The operation of complex conjugation maps this basis to a basis for $(L^2(T) \ominus H^2(T)) \oplus \mathbb{C}$, where $\mathbb{C}$ denotes here the constant functions. Of course the operation of taking the complex conjugate of a symbol of a Toeplitz operator corresponds to taking the adjoint of the Toeplitz operator. Without such algebraic relationships it is doubtful whether the study of classical Toeplitz operators would have enjoyed its present success.

It would seem reasonable to search for a suitable analogy to the classical case as a means of studying the Bergman Toeplitz operators. One is thus led to consider the space $L_h^2(D)$, which is the space of square integrable harmonic functions on the unit disc. $L_h^2(D)$ is a Hilbert space. The proof is similar to the proof that $L_a^2(D)$ is a Hilbert space. $L_h^2(D)$ has a basis consisting of the vectors $\ldots, \sqrt{3}\,\bar{z}^2, \sqrt{2}\,\bar{z}, 1, \sqrt{2}\,z, \sqrt{3}\,z^2, \ldots$. Since $L_a^2(D)$ has the basis $\{1, \sqrt{2}\,z, \sqrt{3}\,z^2, \ldots\}$, the relationship between $L_h^2(D)$ and its subspace $L_a^2(D)$ is much like that between $L^2(T)$ and $H^2(T)$.

Unfortunately the product of two harmonic functions need not be harmonic. One cannot define multiplication operators on $L_h^2(D)$ as one does on $L^2(T)$. There is a remedy for this situation. Let $Q: L^2(D) \to L_h^2(D)$ denote the orthogonal projection. If $\phi \in L_h^\infty(D)$, then define the operator

$$C_\phi: L_h^2(D) \to L_h^2(D)$$

by

$$C_\phi f = Q(\phi f) .$$

C_ϕ is the Bergman space analog of the multiplication operator M_ϕ on $L^2(T)$. If now $P: L_h^2(D) \to L_a^2(D)$ denotes the orthogonal projection, then B can be defined on $L_a^2(D)$ by $B_\phi = PC_\phi$. This way of defining B_ϕ is exactly analogous to the usual definition of T_ϕ in terms of M_ϕ. The hope is of course that spectral properties of the operators C_ϕ are easier to establish than those of the Bergman Toeplitz operators B_ϕ, and that these results will have consequences for the Bergman Toeplitz operators. This work consists of a first study of such operators. Most of the results in this section have proofs which carry over exactly from those in the preceding. In such cases the proofs will not be repeated.

There are many functions $\phi: D \to \mathbb{C}$ for which one can define an operator C_ϕ in a natural fashion. If for example, $\phi(z) = z^{-1}$, then the integral

$$\frac{1}{\pi} \int \phi(z)\, h(z)\, \overline{g(z)}\, dA(z)$$

is well defined for each pair of functions f,g in $L_h^2(D)$. The integral is of course just the inner product $\langle \phi h, g \rangle$ in $L_h^2(D)$. The operator $C_{z^{-1}}: L_h^2(D) \to L_h^2(D)$ defined by this inner product corresponds to the inverse of C_z. The class of functions which can be used to define operators in this manner is large. In this work the desire is to view the operators C_ϕ

as generalizations of the classical multiplication operators on $L^2(T)$. For this reason, attention will be restricted to symbols ϕ, taken from the space $L_h^\infty(D)$. Again an element $\phi \in L_h^\infty(D)$ will not be distinguished from the element of $L^\infty(T)$ which represents the boundary values of ϕ on T.

There is a natural isomorphism $W\colon L^2(T) \to L_h^2(D)$ defined on basis elements by setting

$$W(e^{in\theta}) = \begin{cases} z^n \sqrt{n+1} & \text{if } n \geq 0 , \\ \bar{z}^{|n|} \sqrt{|n|+1} & \text{if } n < 0 . \end{cases}$$

W is of course an extension of the isomorphism $V\colon H^2(T) \to L_a^2(D)$ defined in the previous section. If $\phi \in L_h^\infty(D)$, then a quantitative measure of the difference between M_ϕ and C_ϕ is the operator

$$M_\phi - W^* C_\phi W .$$

The operators $\{C_\phi : \phi \in L_h^\infty(D)\}$ share only some of the algebraic properties of their classical counterpartes $\{M_\phi : \phi \in L^\infty(T)\}$. If ϕ, ϕ_1, ϕ_2 are bounded harmonic functions, then

$$C_{\phi_1} + C_{\phi_2} = C_{\phi_1+\phi_2}$$

and

$$C_\phi^* = C_{\bar{\phi}} .$$

If ϕ_1 is continuous, then the difference

$$C_{\phi_1} C_{\phi_2} - C_{\phi_1\phi_2}$$

is compact, but in general this difference can be very large.

If z represents the independent variable on D, then the operator C_z is the Bergman bilateral shift. A simple computation shows that C_z is a bilateral weighted shift on the natural basis of $L^2_h(D)$ given by

$$C_z(\sqrt{n+1}\ z^n) = \sqrt{\frac{n+1}{n+2}}\,(\sqrt{n+2}\ z^{n+1}) \qquad n \geqq 0 ,$$

$$C_z(\sqrt{n+1}\ \bar{z}^n) = \sqrt{\frac{n}{n+1}}\,(\sqrt{n}\ \bar{z}^{n-1}) \qquad n \geqq 1 .$$

C_z leaves $L^2_a(D)$ invariant and its restriction to this subspace is the Bergman shift B_z. The difference $M_{e^{i\theta}} - W^*C_zW$ is Hilbert-Schmidt but not trace class. C_z is not similar to $M_{e^{i\theta}}$. Indeed, $M_{e^{i\theta}}$ is unitary but the powers C_z^n $(n = 1,2,3,\ldots)$ of C_z converge strongly to zero. C_z and $M_{e^{i\theta}}$ both have the same spectrum, the unit circle T. One interesting difference is that $M_{e^{i\theta}}$ has a logarithm, M_ϕ, where $\phi\colon T \to \mathbb{C}$ is any measurable logarithm of $e^{i\theta}$. C_z doesn't even have a square root. The proof of this fact will be postponed to Section 6.

Proposition 2.1 said that there is a Schur multiplier which maps $\{T_\phi : \phi \in L^\infty(T)\}$ isometrically onto $\{B_\phi : \phi \in L^\infty_h(D)\}$, the correspondence mapping T_ϕ to B_ϕ. The next proposition says that a similar situation holds for $\{M_\phi : \phi \in L^\infty(T)\}$ and $\{C_\phi : \phi \in L^\infty_h(D)\}$. If $\phi \in L^\infty_h(D)$, let $(a_{jk})_{jk\in\mathbb{Z}}$ and $(b_{jk})_{jk\in\mathbb{Z}}$ be the matrices of M_ϕ and C_ϕ with respect to the bases $\{e^{in\theta}\}$ of $L^2(T)$ and $\{\ldots,\sqrt{2}\,\bar{z}, 1, \sqrt{2}\,z,\ldots\}$ of $L^2_h(D)$, respectively. Define $(m_{jk})_{jk\in\mathbb{Z}}$ by

$$m_{jk} = \begin{cases} \sqrt{\dfrac{\min(|j|,|k|) + 1}{\max(|j|,|k|) + 1}} & \text{if } j,k \geqq 0 \text{ or } j,k \leqq 0 , \\[2ex] \sqrt{\dfrac{(|j|+1)(|k|+1)}{(|j| + |k| + 1)^2}} & \text{otherwise} . \end{cases}$$

It should be noted that this definition of m_{jk} is consistent with the definition of Section 2.

3.1 Proposition $(m_{jk})_{jk \in \mathbb{Z}}$ multiplies (in the Schur sense) $\{M_\phi : \phi \in L^\infty(T)\}$ isometrically onto $\{C_\phi : \phi \in L^\infty_h(D)\}$. The correspondence takes M_ϕ to C_ϕ.

Proof: The proof is similar to that of Proposition 2.1 and will be omitted. □

3.2 Proposition $\{C_\phi : \phi \in L^\infty_h(D)\}$ is the WOT closure of the positive powers of C_z and its adjoint $C_{\bar{z}}$.

Proof. See Proposition 2.3. □

3.3 Proposition If $\phi : \bar{D} \to \mathbb{C}$ is continuous on $\bar{D}$ and harmonic on D, then the operator

$$M_\phi - W^* C_\phi W$$

is compact.

Proof. See Proposition 2.4. □

3.4 Theorem If $\phi \in L^\infty_h(D)$, then

$$M_\phi - W^* C_\phi W$$

is Hilbert-Schmidt if and only if the analytic and coanalytic parts of ϕ are in Dirichlet class.

Proof. See Theorem 2.6. □

§4. SOME BESOV SPACES

Besov spaces have a wide range of applications in mathematics. There are many different definitions for Besov spaces living on various domains. For more information on Besov spaces and how they arise see Peetre [6]. In this section a family of Besov spaces of bounded functions on the unit circle is introduced and two equivalent definitions are given. The main space of interest here will be denoted B_1, and is known histroically as the Zygmund class. Of the results given here, at most the characterization of Theorem 4.6 is new. However, published proofs of the other results for the case of the circle seem to be rare, and hence it would seem appropriate to provide them. The approach here was inspired by [8]. Before defining the spaces of interest, it will be necessary to give some notation.

4.1 Definition $L^1(T)$ shall denote the Lebesgue space of integrable functions on the circle T, where T is equipped with normalized arc length measure. If $f \in L^1(T)$, then $\|f\|_1$ shall denote the norm of f in $L^1(T)$. Thus

$$\|f\|_1 = \frac{1}{2\pi}\int_0^{2\pi} |f(e^{i\theta})| \, d\theta \quad .$$

If $f \in L^1(T)$, then the Fourier transform of f is the function $\hat{f}: \mathbb{Z} \to \mathbb{C}$, defined by

$$\hat{f}(n) = \frac{1}{2\pi}\int_0^{2\pi} f(e^{i\theta}) \, e^{-in\theta} \, d\theta \ .$$

The Fourier series of f is the series $\sum_n \hat{f}(n)e^{in\theta}$.

$L^\infty(T)$ shall denote the usual Lebesgue space of bounded measurable functions on T. $\| \ \|_\infty$ shall denote the norm of $L^\infty(T)$. Of course $L^\infty(T) \subset L^1(T)$.

Recall that $H^1 = H^1(T)$ is the collection of elements of $L^1(T)$ whose negative Fourier coefficients vanish and $H^\infty = H^\infty(T)$ is the collection of elements of $L^\infty(T)$ whose negative Fourier coefficients vanish.

If $f, g \in L^1(T)$, define their convolution product $f * g$ to be the element h of $L^1(T)$ defined by

$$h(e^{i\theta}) = \frac{1}{2\pi} \int_0^{2\pi} f(e^{i(\theta-\gamma)})\, g(e^{i\gamma})\, d\gamma .$$

It is well known that this definition makes sense and that $\|f * g\|_1 \leqq \|f\|_1 \|g\|_1$. Furthermore, the convolution product is commutative, and if $f \in L^1(T)$ and $g \in L^\infty(T)$, then $f * g \in L^\infty(T)$ and

$$\|f * g\|_\infty \leqq \|f\|_1 \|g\|_\infty .$$

Also, if $f, g \in L^1(T)$, and if n is any integer, then

$$\widehat{f * g}(n) = \hat{f}(n)\hat{g}(n)$$

<u>4.2 Definition</u> For each positive integer k define $K_k \colon T \to \mathbb{C}$ by

$$K_k(e^{i\theta}) = \sum_{|\ell|<k} \frac{k-|\ell|}{k} e^{i\ell\theta} .$$

K_k is the $(k-1)$ th Fejer kernel on T. K_k is well known to be a positive function on T with $\|K_k\|_1 = 1$.

For each positive integer N, define $W_N \colon T \to \mathbb{C}$ to be the function given by

$$W_N(e^{i\theta}) = (e^{i2^N\theta} + \frac{1}{2}e^{3i2^{N-1}\theta})\, K_{2^{N-1}}(e^{i\theta}) .$$

This is of course a horrible definition. It is instructive to consider $\hat{K}_k$ and $\hat{W}_N$, the Fourier transforms of K_k and W_N, respectively. These are the "tents" pictured in Figure 1. The graph of the Fourier transform of the function $Z_{NM} = W_N + W_{N+1} + \dots + W_{N+M}$ is also pictured in Figure 1. Notice for later, that $\hat{Z}_{NM}(k) \equiv 1$ for all integers k such that $2^N \leqq k \leqq 2^{N+M}$.

Notice also that $\|W_N\|_1 \leq \frac{3}{2}\|K_{2^{N-1}}\|_1 = \frac{3}{2}$.

Finally, as a matter of convenience, define W_0 by $W_0(e^{i\theta}) = 1 + e^{i\theta}$. With this definition, and Z_{NM} as above, it follows that $Z_{0M}(k) = 1$ for $0 \leqq k \leqq 2^M$.

<u>4.3 Definition</u> For $s > 0$, define

$$B_s = \{\phi \in H^1(T) : \sup_{N \in \mathbb{Z}_+} 2^{Ns}\|W_N * \phi\|_\infty < \infty\}$$

For ϕ B_s, define $\|\phi\|_{B_s}$ to be the supremum in the definition above.

<u>4.4 Proposition</u> If $s > 0$, then B_s is a Banach space of continuous functions on the circle T.

<u>Proof</u>. Let $\phi \in B_s$. First it will be shown that ϕ is continuous and $\|\phi\|_\infty \leq \dfrac{\|\phi\|_{B_s}}{1-2^{-s}}$. Indeed, let $P_N = W_N * \phi$ for each positive integer N.

FIGURE 1

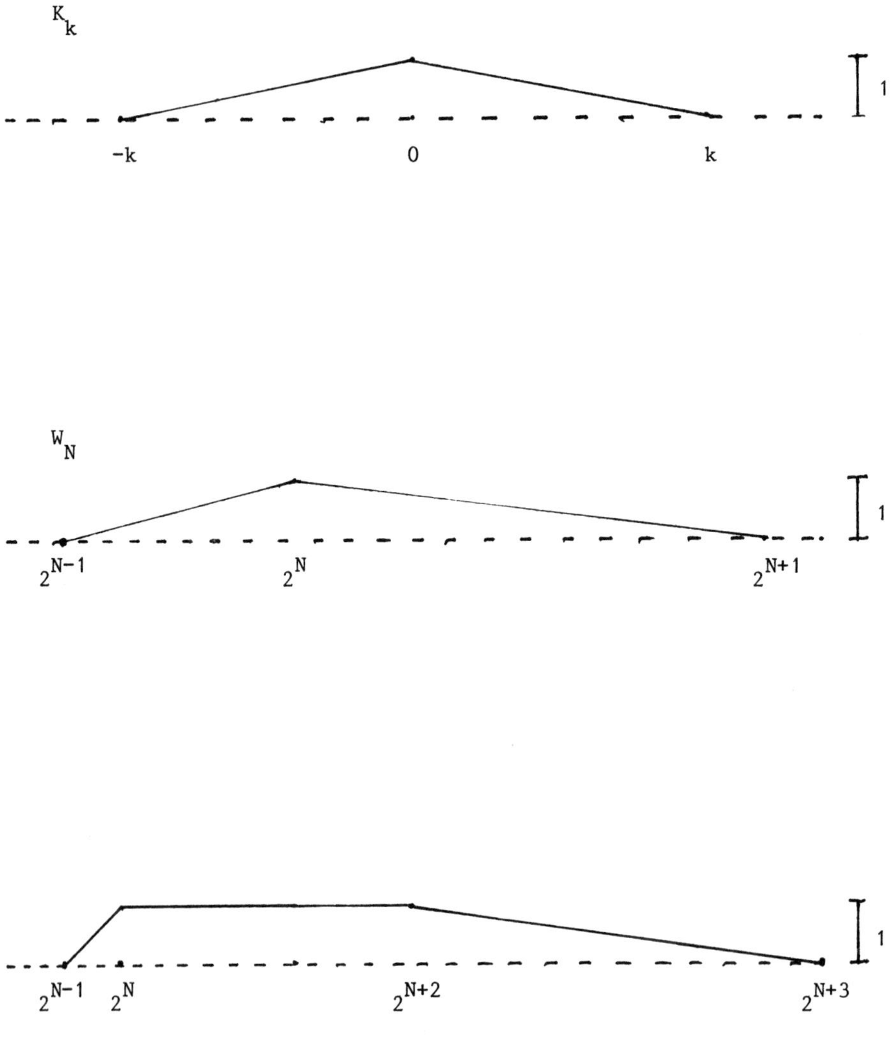

Then P_N is continuous and $\|P_N\|_\infty \leq 2^{-Ns} \|\phi\|_{B_s}$. Clearly $\sum_{N=0}^{\infty} P_N$ converges uniformly to a continuous function $\gamma\colon T \to \mathbb{C}$ and $\|\gamma\|_\infty \leq \frac{\|\phi\|}{1-2^{-s}} B_s$. Convergence in $L^\infty(T)$ implies convergence in $L^1(T)$ so $\hat{\gamma} \in L^1(T)$ and for each integer n,

$$\hat{\gamma}(n) = \sum_{N=0}^{\infty} \hat{P}_N(n)$$

$$= \sum_{N=0}^{\infty} \hat{W}_N(n) \cdot \hat{\phi}(n)$$

$$= \hat{\phi}(n) ,$$

where the last equality is a consequence of the fact that $\sum_{N=0}^{\infty} \hat{W}(n) = 1$ for each positive integer n. Since $\hat{\phi} = \hat{\gamma}$, it follows that $\phi = \gamma$.

Now suppose that $\{\phi_n\}$ is a Cauchy sequence in B_s. By the preceding paragraph, ϕ_n converges in $H^1(T)$ to a function ϕ. It follows that, if $\varepsilon > 0$ is given and $n_0 > 0$ is chosen so that

$$2^{Ns}\|W_N * (\phi_n - \phi_m)\|_\infty < \varepsilon$$

whenever $n,m \geq n_0$ and $N \in \mathbb{Z}_+$, then

$$2^{Ns}\|W_N * (\phi_n - \phi)\|_\infty \leq \varepsilon$$

for each $n \geq n_0$ and $N \in \mathbb{Z}_+$. In other words, ϕ is the limit of $\{\phi_n\}$ in B_s and B_s is a Banach space. □

It should be noted that most definitions of Besov space do not limit candidates to H^1 functions, but allow rather $\phi \in L^1(T)$.

Next, an equivalent formulation of B_s will be given which will be needed in Section 5. First, a definition.

4.5 Definition If $\phi(e^{i\theta}) = \sum_{k=0}^{\infty} b_k e^{ik\theta}$ is a formal power series in $e^{i\theta}$, and if $n > 0$ is a positive integer, then let $T_n = T_n(\phi)$ denote the $n+1$ by $n+1$ matrix

$$T_n = \begin{bmatrix} b_{2n} & & & b_{3n} \\ & & \cdot & \\ & \cdot & & \\ b_{n+1} & \cdot & & \\ b_n & b_{n+1} & & b_{2n} \end{bmatrix}.$$

$\|T_n\|$ shall denote the norm of T_n viewed as an operator on the Hilbert space C^{n+1} (as in Section 2 and 3). If $f,g \in \mathbb{C}^{n+1}$, then $<f,g>$ shall denote the usual inner product on $\mathbb{C}^{n+1}$.

4.6 Theorem If $s > 0$, then $\phi \in B_s$ if and only if $\sup_n n^s \|T_n\| < \infty$.

Proof. Let $\phi(e^{i\theta}) = \sum_{k=0}^{\infty} b_k e^{ik\theta}$ be given. Let $T_n = T_n(\phi)$ be defined as above and assume $K = \sup_n n^s \|T_n\| < \infty$.

Fix an integer $n > 0$ and $\theta \in [0,2\pi]$. Let

$$f(e^{i\theta}) = (W_n * \phi)(e^{i\theta})$$

$$= \sum_{k=2^{n-1}}^{2^n} \frac{k-2^{n-1}}{2^{n-1}} b_k e^{ik\theta} + \sum_{k=2^n+1}^{2^{n+1}} \frac{2^{n+1}-k}{2^n} b_k e^{ik\theta} .$$

It will be shown that $|f(e^{i\theta})| < K(2^s + 2^{-1})2^{-ns}$ for each n and thus that $\phi \in B_s$.

Let $v_1, w_1 \in \mathbb{C}^{2^{n-1}}$ be the vectors

$$v_1 = (1, e^{i\theta}, \ldots , e^{i(2^{n-1}-1)\theta}) ,$$

$$w_1 = (e^{i2^n\theta}, \ldots , e^{i(2^{n-1}+1)\theta}) ;$$

and let M_1 be the 2^{n-1} by 2^{n-1} matrix

$$M_1 = \begin{bmatrix} a_{2^n} & & a_{3\cdot 2^{n-1}-1} \\ & \ddots & \\ a_{2^{n-1}+1} & & a_{2^n} \end{bmatrix} .$$

Notice that M_1 is a submatrix of $T_{2^{n-1}}$. Let

$$f_1(e^{i\theta}) = M_1 v_1 , \bar{w}_1$$

$$= \sum_{\ell=2^{n-1}+1}^{2^n} \sum_{j=0}^{2^{n-1}-1} a_{\ell+j} \, e^{ij\theta} \, e^{i\ell\theta} .$$

Notice that in this double sum, the coefficient $a_{\ell+j}$ always appears paired with $e^{i(\ell+j)\theta}$. Computing the sum is just a matter of computing how many pairs (ℓ,j) sum to a constant value t. If $t = 2^{n-1} + 1$, there is one pair. If $t = 2^{n-1} + 2$, there are two. If $t = 2^n$, there are 2^{n-1}. It is not hard to see that

$$f_1(e^{i\theta}) = \sum_{\ell=2^{n-1}}^{2^n} (\ell-2^{n-1}) e^{i\ell\theta} + \sum_{\ell=2^n+1}^{2^{n+1}} (2^{n+1}-\ell) e^{i\ell\theta} .$$

Letting $\| \|$ denote the Hilbert space norm on $\mathbb{C}^{2^{n-1}}$ as well as the operator norm,

$$|f_1(e^{i\theta})| = |\langle M_1 v_1, \bar{w}_1\rangle|$$

$$\leq \|T_{2^{n-1}}\| \sqrt{2^n}\ \sqrt{2^{n-1}}$$

$$\leq K\, 2^{-(n-1)s}\, 2^{n-1} .$$

Next let $v_2, w_2 \in \mathbb{C}^{2^{n-1}}$ be the vectors

$$v_2 = (1, e^{i\theta}, \ldots , e^{i(2^{n-1}-1)\theta}) ,$$

$$w_2 = (e^{i(2^n+2^{n-1})\theta}, e^{i(2^n+2^{n-1}-1)\theta}, \ldots , e^{i(2^n+1)\theta}) ;$$

M_2 the 2^{n-1} by 2^{n-1} matrix

$$M_2 = \begin{bmatrix} a_{2^n+2^{n-1}} & & a_{2^{n+1}-1} \\ & \ddots & \\ a_{2^n+1} & & a_{2^n+2^{n-1}} \end{bmatrix} ;$$

and $f_2(e^{i\theta}) = \langle M_2 v_2, \bar{w}_2\rangle$.

An entirely analogous computation shows

$$f_2(e^{i\theta}) = \sum_{\ell=2^n}^{2^n+2^{n-1}} (\ell-2^n)e^{i\theta\ell} + \sum_{\ell=2^n+2^{n-1}+1}^{2^{n+1}} (2^{n+1}-\ell)e^{i\ell\theta}$$

and

$$|f_2(e^{i\theta})| \leqq K\, 2^{-ns}\, 2^{n-1} \quad .$$

But

$$f(e^{i\theta}) = \frac{f_1(e^{i\theta})}{2^{n-1}} + \frac{f_2(e^{i\theta})}{2^n} \quad .$$

Therefore $|f(e^{i\theta})| \leqq K(2^s + 2^{-1})2^{-ns}$. Next suppose

$$K = \sup_N \|W_N * \phi\|_\infty\, 2^{Ns} < \infty \quad .$$

Assume $n > 0$ is given. Choose $N \geqq 0$ so that $2^N \leqq n < 2^{N+1}$ and let $Z_N = W_N + W_{N+1} + W_{N+2} + W_{N+3}$. Notice that $\|Z_N * \phi\| \leqq 4K2^{-Ns}$, and that $\widehat{\phi * Z_N}(k) = b_k$ for $2^N \leqq k \leqq 2^{N+3}$. But $[n,3n] \subset [2^N, 2^{N+3}]$. Therefore, the Laurent matrix $M_{Z_N * \phi}$ contains a submatrix of the form

$$M_n = \begin{bmatrix} a_{2n} & & a_n \\ & \ddots & \\ a_{3n} & & a_{2n} \end{bmatrix} \quad .$$

However,

$$\begin{aligned} \|T_n\| &= \|M_n\| \\ &\leqq \|Z_N * \phi\|_\infty \\ &\leqq 4K\, 2^{-ns} \quad . \end{aligned}$$

Therefore $\sup_n \|T_n\|2^{ns} < \infty$. □

If $\phi \in L^\infty(T)$ and if $n > 0$, define the element $\Delta_\eta^2\phi$ of $L^\infty(T)$ by

$$(\Delta_\eta^2 \phi)(e^{i\theta}) = \phi(e^{i(\theta+\eta)}) - 2\phi(e^{i\theta}) + \phi(e^{i(\theta-\eta)}) .$$

This symmetric second difference defines a map from $L^\infty(T)$ onto itself.

<u>4.7 Theorem</u> If $\phi \in H^\infty(T)$, and if $0 < s < 2$, then $\phi \in B_s$ if and only if

$$\sup_{\eta>0} \frac{\|\Delta_\eta^2 \phi\|}{\eta^s} < \infty .$$

The proof will be broken into several lemmas and propositions.

<u>4.8 Lemma</u> If $\phi\colon T \to \mathbb{C}$ is a trigonometric polynomial of the form

$$\phi(e^{i\theta}) = \sum_{n=0}^{N} a_n e^{in\theta} .$$

then

$$\left\| \frac{d}{d\theta} \phi \right\|_\infty \leq N \|\phi\|_\infty .$$

<u>Proof.</u>

$$\left(\frac{d}{d\theta} \phi\right)(e^{i\theta}) = \sum_{n=0}^{N} ia_n n e^{in\theta} .$$

Let

$$J_N(e^{i\theta}) = \sum_{n=0}^{N} in e^{in\theta} + \sum_{n=N+1}^{2N} i(2N-n)e^{in\theta} .$$

J_N is an integrable function defined on the circle T, and

$$\left(\frac{d}{d\theta}\phi\right)(e^{i\theta}) = (J_N * \phi)(e^{i\theta}) .$$

Therefore,

$$\left\|\frac{d}{d\theta}\phi\right\|_\infty \leqq \|J_N\|_1 \; \|\phi\|_\infty .$$

But if K_n denotes as before the (n-1)th Fejer kernel, one has

$$J_N(e^{i\theta}) = iNK_n(e^{i\theta})e^{iN\theta} .$$

Thus, $\|J_N\|_1 = N$ and the desired conclusion follows. □

<u>4.9 Proposition</u> If $\phi = \sum_{n=1}^{\infty} {}_n$, where $\phi_n \in H^1(T)$ has the form

$$\phi_n(e^{i\theta}) = \sum_{k=2^{n-1}}^{2^{n+1}} a_{kn} e^{ik\theta} ,$$

and if $2 > s > 0$ and there is a constant $M > 0$ such that for each integer $n > 0$,

$$\|\phi_n\|_\infty \leqq M2^{-ns} ,$$

then there is a constant $M_s > 0$ so that for each positive n,

$$\|\Delta_\eta^2 \phi\|_\infty < M_s \eta^s .$$

<u>Proof.</u> Let ℓ be chosen so that $2^{-\ell} \leqq \eta < 2^{-\ell+1}$. If $n > \ell$, then $\|\Delta_\eta^2 \phi_n\|_\infty \leqq 4\,\|\phi_n\|_\infty \leqq 4M2^{-ns}$. Therefore

$$\|\Delta_\eta^2 \sum_{n\geqq\ell} \|\phi_n\|_\infty \leqq 4M \sum_{n\geqq\ell} 2^{-ns} \qquad [*]$$

$$= \frac{4M}{1-2^{-s}} 2^{-\ell s} < \frac{3M}{1-2^{-s}} s \; .$$

Next suppose $n < \ell$. Letting the prime symbol " ' " denote differentiation with respect to θ, write

$$\phi_n'(e^{i\theta}) = \phi_n'(e^{i0}) + \theta\gamma(\theta) \; .$$

This is possible because ϕ_n is a polynomial in $e^{i\theta}$. Applying first the Mean Value Theorem and then the previous lemma twice, obtain for $|\theta| < \pi$,

$$|\gamma(\theta)| = \left| \frac{\phi_n'(e^{i\theta}) - \phi_n'(e^{i\theta})}{\theta} \right|$$

$$\leqq \|\phi_n''\|_\infty$$

$$\leqq 2^{n+1} \|\phi_n'\|_\infty$$

$$\leqq 2^{n+1} 2^{n+1} \|\phi_n\|_\infty$$

$$\leqq 2^{2n+2-ns} M \; .$$

Now

$$\phi_n(e^{i\theta}) = \int_0^\theta \phi_n'(e^{it})\, dt + \phi_n(e^{i0})$$

$$= \phi_n(1) + \theta\phi_n'(1) + \int_0^\theta \gamma(t)t\, dt \; .$$

It is easily verified that $\Delta_\eta^2 f \equiv 0$ for linear functions f of θ. Therefore

$$(\Delta_\eta^2 \phi_n)(e^{i\theta}) = \Delta_\eta^2 \left(\int_0^\theta \gamma(t)t\ dt\right)(e^{i\theta})$$

$$= \int_0^\eta \gamma(t)t\ dt - 2\int_0^0 \gamma(t)t\ dt + \int_0^{-\eta} \gamma(t)t\ dt\ .$$

So

$$|(\Delta_\eta^2 \phi_n)(e^{i\theta})| \leqq \int_{-\eta}^{\eta} |\gamma(t)|\,|t|\ dt$$

$$\leqq \|\gamma\|_\infty \int_{-\eta}^{\eta} |t|\ dt$$

$$= \|\gamma\|_\infty\ \eta^2$$

$$\leqq 2^{(2-s)n} \cdot 4M\eta^2\ .$$

By assumption $\eta < 2^{-\ell+1}$ and $0 < s < 2$. Thus,

$$|(\Delta_\eta^2 \sum_{n=1}^{\ell-1} \phi_n)(e^{i\theta})| \leqq \sum_{n=0}^{\ell-1} 2^{(2-s)n}\ 4M2^{-2\ell+2}$$

$$= \frac{2^{(2-s)\ell}-1}{2^{2-s}-1}\ 4M2^{-2\ell+2}$$

$$\leqq \frac{16M}{2^{2-s}-1}\ 2^{-s\ell}$$

$$< \frac{16M}{2^{2-s}-1}\ \eta^s\ .$$

Adding this inequality to * gives

$$|(\Delta_\eta^2 \phi)(e^{i\theta})| \leqq M_s\ \eta^s\ ,$$

where $M_s = \frac{16M}{2^{2-s}-1} + \frac{8M}{1-2^{-s}}$.

By corcular symmetry, the same argument works with θ in place of 0 above. Therefore

$$\|\Delta_\eta^2 \phi\|_\infty \leq M_s \, \eta^s \quad . \qquad \square$$

Remark. This establishes one half of Theorem 4.7.

4.10 Definition If $c, r \in \mathbb{R}$, $r > 0$, define $h(c,r): \mathbb{R} \to \mathbb{C}$ and $h_0(c,r)$ to be the functions whose graphs are pictured below:

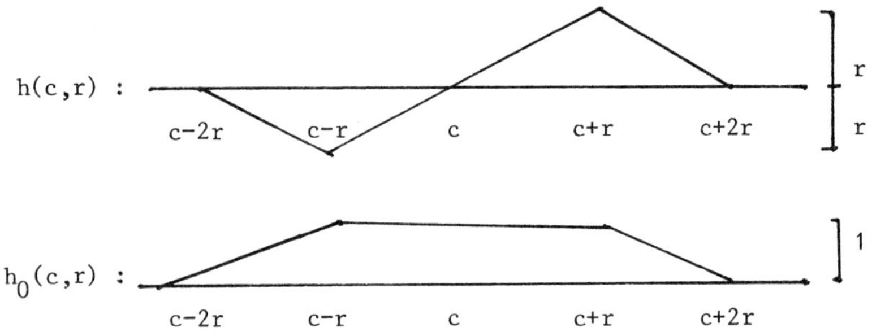

For positive integers n,k define $H(c,r,n,k): T \to \mathbb{C}$ by

$$H(c,r,n,0)(e^{i\theta}) = \sum_j h_0(c,r)(2^{-n}j)e^{ij\theta}$$

$$H_1(e^{i\theta}) = H(c,r,n,1)(e^{i\theta}) = \sum_j h(c,r)(2^{-n}j)e^{ij\theta} \quad ,$$

$$H(c,r,n,k) = \underbrace{H_1 * H_1 * \cdots * H_1}_{k \text{ times}} \quad .$$

4.11 Lemma If $c, r \in \mathbb{R}$, $r > 0$, and if n is a positive integer such that $2^n c$, $2^n r$ are integers, and if k,j are integers, $k > 0$, then

i) $\|H(c,r,n,k)\|_1 \leq 2r^k$;

ii) $\widehat{H(c,r,n,k)}(j) = h^k(c,r)(2^{-n}j)$;

iii) $\|H(c,r,n,0)\|_1 \leq 3$;

iv) $\widehat{H(c,r,n,0)}(j) = h_0(c,r)(2^{-n}j)$.

Proof. (ii) and (iv) are straightforward verifications. To prove (i) let K_j denote the j-1 th Fejer kernel. Notice that $H_1 = H(c,r,n,1)$ can be written in the form $H_1 = rF_+ - rF_-$, where

$$F_+(e^{i\theta}) = e^{i2^n(c+r)\theta} K_{2^n r}(e^{i\theta})$$

and

$$F_-(e^{i\theta}) = e^{i2^n(c-r)\theta} K_{2^n r}(e^{i\theta}) .$$

But $\|F_+\|_1 = \|F_-\|_1 = 1$ and $F_+ * F_- = 0$. Therefore

$$H(c,r,n,k) = H_1 * H_1 * \dots * H_1$$

$$= r^k(\underbrace{F_+ * \dots * F_+}_{k \text{ factors}}) - r^k(\underbrace{F_- * \dots * F_-}_{k \text{ factors}}) ,$$

and the result (i) follows.

(iii) follows from the equality

$$H(c,r,n,0)(e^{i\theta}) = e^{i2^n c\theta} (2K_{2^{n+1}r}(e^{i\theta}) - K_{2^n r}(e^{i\theta})) . \quad \square$$

4.12 The Dilation Lemma. If $g: [a,b] \to \mathbb{C}$ is given by a power series which converges in an open interval containing $[a,b]$, then there is a constant M and an extension $\bar{g}: \mathbb{R} \to \mathbb{C}$ of g such that if $G_n: T \to \mathbb{C}$ is defined by

$$G_n(e^{i\theta}) = \sum_{k=-\infty}^{\infty} \bar{g}(2^{-n}k)e^{ik\theta} ,$$

then

$$\|G_n\|_{L^1(T)} \leq M$$

for each positive integer n.

Proof. By expanding the interval $[a,b]$ slightly, it may be assumed that a,b, and $c = \frac{a+b}{2}$ are all dyadic numbers. Let $r = \frac{b-a}{2}$. Write $g(x) = \sum_{k=0}^{\infty} a_k(x-c)^k$, $x \in [a,b]$. This series is known to converge absolutely in this interval so that $\sum |a_k| r^k < \infty$. Let $h_0 = h_0(c,r): \mathbb{R} \to \mathbb{C}$ and $h = h(c,r): \mathbb{R} \to \mathbb{C}$ be as in Definition 4.10. Now define $\tilde{g}: \mathbb{R} \to \mathbb{C}$ by

$$\tilde{g}(x) = a_0 h_0(x) + \sum_{k=1}^{\infty} a_k\, h^k(x) .$$

Since $\|h\|_\infty = r$, this series converges uniformly. On $[a,b]$, $h_0(x) = 1$ and $h(x) = x-c$ so $\tilde{g}$ is an extension of g.

Let n be any positive integer such that $2^n c$ and $2^n r$ are integers. For each positive integer k let $H_k = H(c,r,n,k)$ be as in Definition 4.10. By Lemma 4.11

$$\sum |a_k|\, \|H_k\|_1 \leq \sum 3|a_k|\, r^k = M < \infty .$$

Notice that M is independent of n. Therefore the series

$$F_n = \sum_{k=0}^{\infty} a_k H_k$$

defines an element F of $L^1(T)$ and $\|F_n\|_1 \leq M$. It remains to show that $F_n = G_n$, where G_n is defined above. To show this it suffices to show that $\hat{F}_n(k) = \tilde{g}(2^{-n}k)$ for each integer k. However, by Lemma 4.11,

$$\hat{F}_n(k) = \sum_{\ell=0}^{\infty} a_\ell \hat{H}_\ell(k)$$

$$= a_0 h_0(2^{-n}k) + \sum_{\ell=1}^{\infty} a_\ell h^\ell(2^{-n}k)$$

$$= \tilde{g}(2^{-n}k) ,$$

where the first equality follows by the $L^1(T)$ summability of the series defining F_n.

This shows that there is an absolute constant M so that $\|G_n\|_1 \leq M$ whenever $2^n c$ and $2^n r$ are integers. Since c and r are dyadic, this happens for all but finitely many choices of positive integers n. $\tilde{g}$ has compact support. Thus $G_n \in L^1(T)$ for the remaining values of n and the lemma follows. □

<u>4.13 Definition</u> Let $g: [\frac{1}{2},2] \to \mathbb{R}$ be the function $g(x) = \frac{1}{2}(\cos(x)-1)^{-1}$. For each integer n define $G_n: T \to \mathbb{C}$ by

$$G_n(e^{i\theta}) = \sum_{k=2^{n-1}}^{2^{n+1}} g(2^{-n}k)e^{ik\theta} .$$

4.14 Lemma If W_n, G_n are as in Definition 4.2 and 4.13, and if $\phi \in L^\infty(T)$, then

$$W_n * \phi = G_n * W_n * (\Delta^2_{2^{-n}} \phi) \ .$$

Proof. By a change of variables it follows that

$$W_n * (\Delta^2_{2^{-n}} \phi) = \Delta^2_{2^{-n}} (W_n * \phi) \ .$$

$W_n * \phi$ is a polynomial of the form

$$\sum_{k=2^{n-1}}^{2^{n+1}} a_k e^{ik\theta} \ ,$$

so it suffices to show that

$$\phi = G_n * (\Delta^2_{2^{-n}} \phi) \ ,$$

where $\phi(e^{i\theta}) = e^{ik\theta}$, $2^{n-1} \leqq k \leqq 2^{n+1}$.

$$\begin{aligned}(\Delta^2_{2^{-n}} \phi)(e^{i\theta}) &= e^{i(\theta+2^{-n})k} - 2e^{i\theta k} + e^{i(\theta-2^{-n})k} \\ &= (e^{i2^{-n}k} - 2 + e^{-i2^{-n}k})e^{i\theta k} \\ &= 2(\cos(2^{-n}k) - 1)e^{i\theta k} \ .\end{aligned}$$

Thus,

$$\begin{aligned}(G_n * (\Delta^2_{2^{-n}} \phi))(e^{i\theta}) &= g(2^{-n}k)2(\cos(2^{-n}k)-1)e^{i\theta k} \\ &= e^{ik\theta} \\ &= \phi \ . \quad \square\end{aligned}$$

4.15 Theorem There is a constant M such that if $\phi \in L^\infty(T)$ and n is a positive integer, then

$$\|W_n * \phi\|_\infty \leq M \|\Delta^2_{2^{-n}} \phi\|_\infty .$$

Proof. With the notation of the preceding lemma and definition,

$$\|W_n * \phi\|_\infty = \|G_n * W_n * (\Delta^2_{2^{-n}} \phi)\|_\infty$$

$$\leq \|G_n * W_n\|_1 \|\Delta^2_{2^{-n}} \phi\|_\infty .$$

Recall that W_n has the form

$$W_n(e^{i\theta}) = \sum_{k=2^{n-1}}^{2^{n+1}} w_k e^{ik\theta} .$$

Apply the dilation lemma to $g\colon [\frac{1}{2},2] \to \mathbb{C}$ and get an extension $\tilde{g}\colon \mathbb{R}\to$ of g, and a constant M, so that if $\tilde{G}_n\colon T \to \mathbb{C}$ is defined by

$$\tilde{G}_n(e^{i\theta}) = \sum_k \hat{g}(2^{-n}k)e^{ik\theta} ,$$

then $\|G_n\| \leq M$. For $2^{n-1} \leq k \leq 2^{n+1}$,

$$\tilde{g}(2^{-n}k) = g(2^{-n}k)$$

since $\frac{1}{2} \leq 2^{-n}k \leq 2$. Therefore $\tilde{G}_n * W_n = G_n * W_n$. Hence,

$$\|\tilde{G}_n * W_n\|_1 = \|\tilde{G}_n * W_n\|_1$$

$$\leq \|G_n\|_1 \|W_n\|_1$$

$$\leq M \cdot 2. \quad \square$$

Theorem 4.15 together with Proposition 4.9 now suffices to prove Theorem 4.7.

The following proposition determines the closure of the polynomials in B_s for $0 < s < 2$.

4.16 Definition If $0 < s < 2$, define $f_i\colon B_s \to [0,\infty)$ $(i=1,2,3)$ by

$$f_1(\phi) = \limsup_{n\to\infty} n^s \|T_n(\phi)\| ,$$

$$f_2(\phi) = \limsup_{n\to\infty} 2^{ns} \|W_n * \phi\|_\infty ,$$

$$f_3(\phi) = \limsup_{t\to 0^+} \frac{\|\Delta_t^2 \phi\|_\infty}{t^s} .$$

Define $S_i^s = \{\phi \in B_s : f_i(\phi) = 0\}$.

4.17 Proposition If $0 < s < 2$ and $i \in \{1,2,3\}$, and if P_s is the closure of the polynomials in B_s, then

$$P_s = S_i^s .$$

Proof. First notice that the functions $f_i\colon B_2 \to [0,\infty)$ are well defined and continuous. Thus S_i^s $(i=1,2,3)$ is closed. If ϕ is a polynomial, then $f_i(\phi)$ is clearly zero for each choice of i. Thus $P_s \subset S_i^s$ for each i. By Theorem 4.15, there is a constant $M > 0$ so that $\|W_n * \phi\|_\infty \leqq M\|\Delta_{2^{-n}}^2 \phi\|$ for each integer n. Thus $S_3^s \subset S_2^s$. If $n > 0$ and N is chosen so that $2^N \leqq n < 2^{N+1}$, then the proof of Theorem 4.6 shows that

$$\|T_n\| \leqq \|(W_N + W_{N+1} + W_{N+2} + W_{N+3}) * \phi\|$$

and

$$\|W_N * \phi\|_\infty \leqq \|T_{2^{N-1}}\| + \|T_{2^N}\| \quad .$$

This shows that $S_1^s = S_2^s$. It remains only to show that $S_2^s \subset P_s$.

If $\phi \in S_2^s$ and $\varepsilon > 0$, then there is an integer N_0 such that

$$\|W_N * \phi\|_\infty < \frac{\varepsilon}{2}\, 2^{-Ns}$$

whenever $N \geqq N_0$. If K_J denotes the $(J-1)$th Fejer kernel, then $\|K_J\|_{L^1(T)} \equiv 1$, and if g is any polynomial, then

$$\lim_{J\to\infty} \|K_J * g - g\|_\infty = 0 \quad .$$

Now $W_N * \phi$ is a polynomial for each N, so choose J so that

$$\|K_J * W_N * \phi - W_N * \phi\| < \varepsilon 2^{N_0 s}$$

for $N = 0,1,2,\ldots,N_0$ and let $\phi_J = K_J * \phi$. Then ϕ_J is a polynomial and

$$\|W_N * (\phi-\phi_J)\|_\infty < \varepsilon 2^{-N_0 s} \leqq \varepsilon 2^{-Ns}$$

if $N = 0,1,\ldots,N_0$, and

$$\|W_N * (\phi-\phi_J)\|_\infty \leqq \|W_N * \phi\|_\infty + \|W_N * K_J * \phi\|_\infty \quad ,$$

if $N \geqq N_0$. But

$$\begin{aligned}\|W_N * K_J * \phi\|_\infty &= \|K_J * (W_N * \phi)\|_\infty \\ &\leqq \|K_J\|_1 \, \|W_N * \phi\|_\infty \\ &\leqq \|W_N * \phi\|_\infty \quad .\end{aligned}$$

Therefore

$$\|W_N * (\phi-\phi_J)\|_\infty \leqq 2\,\|W_N * \phi\|_\infty$$

$$\leqq \varepsilon 2^{-Ns} ,$$

if $N \geqq N_0$.

Clearly ϕ is in the closure of the polynomials in B_s and so $S_2^s \subset P_s$. □

§5. THE COMMUTANT OF THE BERGMAN BILATERAL SHIFT

This chapter is devoted to determining the commutant of the Bergman bilateral shift C_z. Recall that the commutant of an operator T, on a Hilbert space $\mathcal{H}$, is the collection $\{T\}'$ of operators which commute with T. $\{T\}'$ is a strongly closed algebra in $B(\mathcal{H})$.

The commutant of the Bergman shift B_z has the description

$$\{B_z\}' = \{B_\phi : \phi \in L_a^\infty(D)\} \quad .$$

To see this suppose $T \in \{B_z\}'$. Let $\phi = T1 \in L_a^2(D)$. If n is a positive integer, then

$$Tz^n = TB_z^n 1 = B_z^n T1 = B_z^n \phi = z^n \phi \ .$$

Hence, on the polynomials, T is multiplication by the analytic function ϕ. Using the fact that the polynomials are dense in $L_a^2(D)$, one can show that the analytic function ϕ must also be bounded and that $T = B_\phi$. For the complete details see Conway [2], p. 147.

For the unilateral shift $T_{e^{i\theta}}$, the commutant has a similar description

$$\{T_{e^{i\theta}}\}' = \{T_\phi : \phi \in H^\infty(\mathbb{T})\} \quad .$$

$H^\infty(\mathbb{T})$ is the collection of bounded functions on $\mathbb{T}$ whose negative Fourier coefficients vanish. Via the Poisson integral there is the identification $H^\infty(\mathbb{T}) \cong L_a^\infty(D)$.

The bilateral shift $M_{e^{i\theta}}$ has the commutant

$$\{M_{e^{i\theta}}\}' = \{M_\phi : \phi \in L^\infty(T)\} \quad .$$

One would expect an analogous result for the Bergman bilateral shift. In fact, the commutant of C_z can be explicitly determined as an algebra of functions on the circle, but it is not the algebra $L_h^\infty(D)$. The main result of this section is that the commutant of C_z can be identified with an algebra A of the form $A = H^\infty \oplus \bar{B}$, where $H^\infty = H^\infty(T)$ and $\bar{B}$ denotes the algebra of functions whose complex conjugates are in the Zygmund class. The multiplication in A corresponds to the pointwise multiplication of bounded functions on the circle. The proof of this requires first a development of some of the machinery in the survey article "Weighted shift operators and analytic function theory" by Allen Shields [5]. First the necessary material from Shield's article will be given, though without proofs.

An operator S on a Hilbert space H is a bilateral weighted shift if there is a doubly infinite sequence of constants $\{w_n\}_{n\in\mathbb{Z}}$, and an orthonormal basis $\{f_n\}_{n\in\mathbb{Z}}$ for H such that for each integer n,

$$Sf_n = w_n f_{n+1} \quad .$$

By choosing the basis appropriately, the weights may and will all be taken to be nonnegative.

The classical bilateral shift, $M_{e^{i\theta}}$ on $L^2(T)$, is a shift with weight sequence $w_n \equiv 1$ and orthonormal "shift" basis $\{f_n\}$ given by $f_n = e^{in\theta}$. The Bergman bilateral shift has weight sequence $\{w_n\}$ defined by

$$w_n = \begin{cases} \sqrt{\dfrac{n+1}{n+2}} & \text{if } n = 0,1,2,\dots \ , \\[2ex] \sqrt{\dfrac{|n|}{|n|+1}} & \text{if } n = -1,-2,\dots \ . \end{cases}$$

C_z shifts the basis f_n where

$$f_n = \begin{cases} z^n \sqrt{n+1} & \text{if } n = 0,1,2,\ldots , \\ \bar{z}^{|n|} \sqrt{|n|+1} & \text{if } n = -1,-2,\ldots . \end{cases}$$

If S is a fixed injective bilateral weighted shift with positive weight sequence $\{w_n\}$ and shift basis $\{f_n\}$, then define a sequence $\{\beta(n)\}_{n\in\mathbb{Z}}$ by

$$\beta(n) = \begin{cases} 1 & \text{if } n = 0 , \\ w_0 \cdot \ldots \cdot w_{n-1} & \text{if } n > 0 , \\ (w_{-1} w_{-2} \cdots w_n)^{-1} & \text{if } n < 0 . \end{cases}$$

Here the weights are assumed to be strictly positive. Of course the sequence $\beta(n)$ is uniquely determined by the sequence $\{w_n\}$, and conversely, given the sequence $\{\beta(n)\}$, one can compute the sequence $\{w_n\}$ whence it came. In fact, $w_n = {}^{\beta(n+1)}/\beta(n)$ for each integer n.

Let $L^2(\beta)$ denote the Hilbert space of formal series of the form $f \sim \sum_{n\in\mathbb{Z}} a_n e_n$, where $a_n \in \mathbb{C}$ for each integer n and where

$$\|f\|_{L^2(\beta)} = \Big(\sum_{n\in\mathbb{Z}} |a_n|^2 \beta(n)^2\Big)^{1/2} < \infty .$$

The vectors e_n may be viewed as formal coordinates or as powers of $e^{i\theta}$.

On $L^2(\beta)$ define the operator $\tilde{M}_{e_1} : L^2(\beta) \to L^2(\beta)$ by

$$\tilde{M}_{e_1}\Big(\sum a_n e_n\Big) = \sum a_n e_{n+1} .$$

$\tilde{M}_{e_1}$ is a shift on the orthonormal basis, $\{e_n/\beta(n)\}$ of $L^2(\beta)$ with

$$\tilde{M}_{e_1}\left(\frac{e_n}{\beta(n)}\right) = \left(\frac{\beta(n+1)}{\beta(n)}\right)\left(\frac{e_{n+1}}{\beta(n+1)}\right) .$$

Thus the weight sequences of the shifts S and M_{e_1} are identical. It follows that M_{e_1} on $L^2(\beta)$ is unitarily equivalent to S on $\mathcal{H}$.

For the bilateral Bergman shift, the sequence $\{\beta(n)\}$ is defined by

$$\beta(n) = \begin{cases} (n+1)^{-1/2} & \text{if } n = 0,1,2,\ldots \\ (|n|+1)^{1/2} & \text{if } n = -1,-2,\ldots \end{cases} .$$

The elements of $L^2(\beta)$ may be thought of as formal Laurent series. If $\phi \sim \sum b_n e_n$ is a formal Laurent series, and if $f \sim \sum a_n e_n \in L^2(\beta)$, then one can attempt to define a formal product ϕf by

$$\phi f \sim \sum c_n e_n ,$$

where

$$c_n = \sum_{k \in \mathbb{Z}} a_{n-k} b_k .$$

Of course the series $\sum_k a_{n-k} b_k$ need not always be summable, and even if they are, the Laurent series $\sum c_n e_n$ need not corresponds to an element of $L^2(\beta)$. For some choices of ϕ, however, the product ϕf defines an element of $L^2(\beta)$ for each $f \in L^2(\beta)$. Whenever this is the case the Closed Graph Theorem implies that the operator $f \to \phi f$ is bounded on $L^2(\beta)$. If ϕ has the form $\phi \sim \sum_{n=-N}^{N} b_n e_n$, then this is certainly the case. The operator $\phi \to \phi f$ so defined on $L^2(\beta)$ will be denoted by $\tilde{M}_\phi$. Notice that this notation is consistent with the definition of $\tilde{M}_{e_1}$ given earlier. The collection of ϕ for which $\tilde{M}_\phi : L^2(\beta) \to L^2(\beta)$ defines a bounded operator will be denoted by $L^\infty(\beta)$. For $\phi \in L^\infty(\beta)$ define $\|\phi\|_{L^\infty(\beta)} = \|\tilde{M}_\phi\|$, where $\|\tilde{M}_\phi\|$ is the operator norm. This norm turns $L^\infty(\beta)$ into a Banach algebra. If $\phi_1, \phi_2 \in L^\infty(\beta)$, then there is a $\phi_3 \in L^\infty(\beta)$ so that $\tilde{M}_{\phi_1} \tilde{M}_{\phi_2} = \tilde{M}_{\phi_3}$. This is the multiplication on $L^\infty(\beta)$ and corresponds to formal Laurent series multiplication.

The following theorem gives the connection between the commutant of S and the algebra $L^\infty(\beta)$. The notation of the theorem will be that given above. The proof is in Shields [5], p. 62.

5.1 Theorem If S is an injective bilateral weighted shift on the Hilbert space $\mathcal{H}$, and if $U: \mathcal{H} \to L^2(\beta)$ is the isomorphism defined by $U(f_n) = e_n/\beta(n)$ $(n \in \mathbb{Z})$, then there is an isometric identification $I: L^\infty(\beta) \to \{S\}'$ of the commutant of S with the algebra $L^\infty(\beta)$ defined by

$$I(\phi) = U^* \tilde{M}_\phi U . \qquad \square$$

The preceding theorem gives the form of the commutant of a shift, but is not very enlightening as to the specifics of which formal Laurent series belong to $L^\infty(\beta)$. For a given shift, the task is to give a concrete realization of the algebra $L^\infty(\beta)$. In the following sequence of propositions, an identification of $L^\infty(\beta)$ will be given for the case of the Bergman bilateral shift. The realization will take the form of the algebra of functions alluded to at the beginning of this section. The sequence $\{\beta(n)\}$ will be fixed to be the sequence defined earlier which corresponds to the Bergman bilateral shift C_z. The techniques generalize to certain other choices for $\{\beta(n)\}$ and comments about this will be made in Section 7.

The techniques of this section involve matrix methods. First notation must be developed. It will be useful to view finite and infinite dimensional matrices as operators on Hilbert space. A matrix is bounded exactly when the operator it corresponds to is bounded. Matrices will always be assumed to represent operators with respect to orthonormal bases. With this assumption, there can be no ambiguity. Thus if M is an $n\times n$ matrix, M can be viewed as an operator on the Hilbert space $\mathbb{C}^n$. Then $\|M\|$ shall denote the operator norm of M. If M is an infinite dimensional matrix, then $\|M\| < \infty$ indicates that the operator corresponding to the matrix M is bounded with operator norm $\|M\|$. $\|M\| = \infty$ simply means that M is not the matrix of a bounded operator with respect to an orthonormal basis.

If $\phi \sim \sum a_n e_n$ is any formal Laurent series ($a_n \in \mathbb{C}$ for $n \in \mathbb{Z}$), then $\tilde{M}_\phi$ shall denote the matrix

$$(b_{jk})_{jk\in\mathbb{Z}} ,$$

where

$$b_{jk} = \frac{\beta(j)}{\beta(k)} a_{j-k} .$$

If $\phi \in L^\infty(\beta)$, then $\tilde{M}_\phi$ is the matrix of the bounded operator of formal multiplication by ϕ on $L^2(\beta)$ expressed with respect to the orthonormal basis $\{\frac{e_n}{\beta(n)} : n \in \mathbb{Z}\}$. In fact, $\tilde{M}_\phi$ corresponds to the matrix of a bounded operator on $L^2(\beta)$ exactly when $\phi \in L^\infty(\beta)$. Define also

$$\|\phi\|_{L^\infty(\phi)} = \|\tilde{M}_\phi\| .$$

It will be useful to be able to regard $L^\infty(T)$ as a space of formal Laurent series. Thus if $\phi \sim \sum a_n e_n$ is a formal Laurent series, then the statement $\phi \in L^\infty(T)$ shall mean that the series $\sum a_n e^{in\theta}$ is the Fourier series of a bounded function on T. $\|\phi\|_\infty$ shall denote the bound of this function. M_ϕ shall denote the matrix $(a_{jk})_{jk\in\mathbb{Z}}$ given by $a_{jk} = a_{j-k}$. It is well known that $\phi \in L^\infty(T)$ exactly whenever $\|M_\phi\| < \infty$ and then $\|\phi\|_\infty = M_\phi$.

If a and b are integers, $a \leq b$, then $[a,b]$ denotes the set $\{a, a+1, \ldots, b\}$. A subaset $R \subset \mathbb{Z} \times \mathbb{Z}$ of the form $R = [x_1, x_2] \times [y_1, y_2]$ is called a rectangle. If M is the infinite matrix $(c_{jk})_{jk \in \mathbb{Z}}$, then the R-block of M is the submatrix $M(R)$ defined by

$$M(R) = \begin{bmatrix} c_{x_1 y_1} & & c_{x_1 y_2} \\ & \ddots & \\ c_{x_2 y_1} & & c_{x_2 y_2} \end{bmatrix}$$

The norm of a matrix M is, of course, just the supremum of the norms of its finite dimensional submatrices.

The next lemma is a key to all the computations which follow. There is much symmetry lurking within the matrix $\tilde{M}_\phi$ and this lemma exploits it.

5.2 <u>The Factorization Lemma</u>. If $\phi \sim \sum a_n e_n$ is a formal Laurent series and if $R = [x_1, x_2] \times [y_1, y_2]$ is a rectangle, then $\tilde{M}_\phi(R)$ factors into a product of the form

$$\tilde{M}_\phi(R) = \begin{bmatrix} \beta(x_1) & & 0 \\ & \ddots & \\ 0 & & \beta(x_2) \end{bmatrix} M_\phi(R) \begin{bmatrix} \beta(y_1) & & \\ & \ddots & \\ & & \beta(y_2) \end{bmatrix}^{-1} .$$

Moreover $\dfrac{\beta(x_2)}{\beta(y_1)} \|M_\phi(R)\| \leq \|\tilde{M}_\phi(R)\| \leq \dfrac{\beta(x_1)}{\beta(y_2)} \|M_\phi(R)\|$.

<u>Proof</u>. Recall that $\tilde{\tilde{M}}_\phi$ has the form $(b_{jk})_{jk \in \mathbb{Z}}$, where $b_{jk} = \dfrac{\beta(j)}{\beta(k)} a_{j-k}$ and M_ϕ has the form $(a_{jk})_{jk \in \mathbb{Z}}$, where $a_{jk} = a_{j-k}$. The factorization is now obvious. To get the norm estimates, recall that from general

principles if $\tilde{M} = AMB^{-1}$, where A and B are invertible, then $\|\tilde{M}\| \leq \|A\| \, \|M\| \, \|B^{-1}\|$ and $\|A^{-1}\| \, \|\tilde{M}\| \, \|B\| \geq \|M\|$. Applying these estimates and the fact that the sequence $\{\beta(n)\}$ is a positive monotone decreasing sequence gives the estimates above. □

5.3 Lemma $L^\infty(\beta) \subset L^\infty(T)$. Moreover, if $\phi \in L^\infty(\beta)$, then $\|\phi\|_{L^\infty(T)} \leq \|\phi\|_{L^\infty(\beta)}$.

Proof. If $\phi \sim \sum a_n e_n \in L^\infty(\beta)$ and $n > 0$, then let M_n be the matrix

$$M_n = \begin{bmatrix} a_0 & a_{-1} & & & a_{-n} \\ a_1 & a_0 & & & \\ & & \ddots & & \\ & & & a_0 & a_{-1} \\ a_n & & & a_1 & a_0 \end{bmatrix}.$$

To show $\phi \in L^\infty(T)$, if suffices to show that $\sup_{n>0} \|M_n\| < \infty$. For each positive integer m, let R_m be the rectangle $R_m = [m,m+n] \times [m,m+n]$. Notice that $M_\phi(R_m) = M_n$. Thus by the Factorization Lemma,

$$\|M_n\| \leq \frac{\beta(m+n)}{\beta(m)} \|\tilde{M}_\phi(R_m)\|$$

$$= \sqrt{\frac{m+n+1}{m+1}} \, \|\tilde{M}_\phi(R_m)\| .$$

The inequality above is valid for all $m > 0$ and $\|\tilde{M}_\phi(R_m)\| \leq \|\tilde{M}_\phi\|$ for all m. Clearly

$$\|M_n\| \leqq \|\tilde{M}_\phi\|$$

for all n. Hence $\phi \in L^\infty(T)$ and $\|\phi\|_\infty \leqq \|\phi\|_{L^\infty(\beta)}$. □

The following fact will be needed for the proof of Theorem 5.5.

5.4 Lemma If ℓ, m, n are positive integers and if T and F are matrices of the form

$$T = \begin{bmatrix} a_{2n} & & a_{3n} \\ & \ddots & \\ a_n & & a_{2n} \end{bmatrix},$$

$$F = \begin{bmatrix} a_{m+n} & & a_{m+n+\ell} \\ & \ddots & \\ a_n & & a_{n+\ell} \end{bmatrix},$$

where T and F are constant along diagonals, and if $3n \geqq n+m+\ell$, then

$$\|F\| \leqq 2\|T\| .$$

Proof. If $n \geqq m$ and $n \geqq \ell$, then F is a submatrix of T and $\|F\| \leqq \|T\|$. Assume $m > n$. Then $\ell < n$ and the matrix F admits a decomposition into two submatrices A and B, where

$$A = \begin{bmatrix} a_{2n} & & a_{2n+\ell} \\ & \ddots & \\ a_n & & a_{n+\ell} \end{bmatrix},$$

$$B = \begin{bmatrix} a_{n+m} & & a_{n+m+\ell} \\ & \ddots & \\ a_{2n+1} & & a_{2n+\ell+1} \end{bmatrix} .$$

A is clearly a submatrix of T. Consider the submatrix C of T given by

$$C = \begin{bmatrix} a_{3n-\ell} & a_{3n} \\ & \\ a_{2n+1} & a_{2n+\ell+1} \end{bmatrix} .$$

Since $3n-\ell > n+m$ and $3n > n+m+\ell$, B is seen to be a submatrix of C, hence also of T. Therefore $\|F\| \leqq \|A\| + \|B\| \leqq 2\,\|T\|$. □

<u>5.5 Theorem</u> If $\phi \sim \sum a_n e_n$ is a formal Laurent series, and if T_n is the matrix defined by

$$T_n = \begin{bmatrix} a_{-2n} & & a_{-3n} \\ & \ddots & \\ a_{-n-1} & & \\ a_{-n} & a_{-n-1} & a_{-2n} \end{bmatrix} ,$$

then $\phi \in L^{\infty}(\beta)$ if and only if $\phi \in L^{\infty}(T)$ and $\sup_{n>0} n\,\|T_n\| < \infty$. Moreover, $\|\phi\|_{\infty} + \sup_{n>0} n\,\|T_n\|$ is an equivalent norm for $\|\tilde{M}_{\phi}\|$.

<u>Proof</u>. First assume $\phi \in L^{\infty}(\beta)$. By Lemma 5.3, $\phi \in L^{\infty}(T)$. Let n be a positive integer. For convenience assume n is even. If R is the rectangle $R = [\frac{-3n}{2}, \frac{-n}{2}] \times [\frac{n}{2}, \frac{3n}{2}]$, then $M_{\phi}(R) = T_n$ and the Factorization Lemma gives that

$$\frac{\beta(-\frac{n}{2})}{\beta(\frac{n}{2})} \|T_n\| \leqq \|\tilde{M}_\phi(R)\| \leqq \|\tilde{M}_\phi\| \quad .$$

But $\dfrac{\beta(-\frac{n}{2})}{\beta(\frac{n}{2})} = \sqrt{\frac{n}{2}+1}\sqrt{\frac{n}{2}+1} = \frac{n}{2} + 1$. Therefore

$$\sup_n n \|T_n\| \leqq 2 \|\tilde{M}_\phi\| \quad .$$

This completes one half of the proof. Before proceeding it will be necessary to establish some more notation.

If $M = (c_{jk})_{jk\in\mathbb{Z}}$ is a matrix and $S \subset \mathbb{Z} \times \mathbb{Z}$, then define $M(S)$ to be the matrix $(d_{jk})_{jk\in\mathbb{Z}}$ given by

$$d_{jk} = \begin{cases} c_{jk} & \text{if } (j,k) \in S \\ 0 & \text{if } (j,k) \notin S \quad . \end{cases}$$

If R is a rectangle, then this notation disagrees only formally with the previous definition of $M(R)$. With the obvious modification the Factorization Lemma still applies and the norm estimates of that lemma remain unchanged.

Two rectangles $R_i = [x_{i1},x_{i2}] \times [y_{i1},y_{i2}]$ $(i = 1,2)$ are said to be orthogonal if the sets $[x_{11},x_{12}] \cap [x_{21},x_{22}]$ and $[y_{11},y_{12}] \cap [y_{21},y_{22}]$ are each empty. If $R_1, R_2, R_3, \ldots$ is a sequence of mutually orthogonal rectangles, then

$$\|M(\bigcup_i R_i)\| = \sup_i \|M(R_i)\| \quad .$$

The proof of this is identical to the proof that the norm of a diagonal matrix is just the supremum of the absolute values of the diagonal entries.

Now the proof of the theorem can be completed. Assume therefore that $\phi \in L^{\infty}(T)$ and $K = \sup n \|T_n\| < \infty$. To prove that $\tilde{M}_\phi$ is bounded, it is convenient to argue that each of the four submatrices $\tilde{M}_{++} = \tilde{M}_\phi(\mathbb{Z}_+ \times \mathbb{Z}_+)$, $\tilde{M}_{+-} = \tilde{M}_\phi(\mathbb{Z}_+ \times \mathbb{Z}_-)$, $\tilde{M}_{-+} = \tilde{M}_\phi(\mathbb{Z}_- \times \mathbb{Z}_+)$, and $\tilde{M}_{--} = \tilde{M}_\phi(\mathbb{Z}_- \times \mathbb{Z}_-)$ is bounded.

Case 1: "$\tilde{M}_{-+}$ is bounded." This is the heart of the theorem.

Partition $\mathbb{Z}_- \times \mathbb{Z}_+$ into rectangles R_{jk} $(j,k = 0,1,2,\ldots)$, where $R_{jk} = [-2^{j+1} + 1, -2^j] \times [2^k, 2^{k+1} - 1]$. This partition is illustrated in Figure 2. For each integer n let $D_n = \cup_{j-k=n} R_{jk}$. D_n may be thought of as a sort of diagonal. Notice that the rectangles which make up any given diagonal are mutually orthogonal. Assume $n \geqq 0$. Then

$$\|\tilde{M}_\phi(D_n)\| = \sup_{j-k=n} \|\tilde{M}_\phi(R_{jk})\|$$

$$= \sup_{k \geqq 0} \|\tilde{M}_\phi(R_{k+n,k})\| \quad .$$

By the Factorization Lemma,

$$\|\tilde{M}_\phi(R_{k+n,k})\| \leqq \frac{\beta(-2^{k+n+1} + 1)}{\beta(2^{k+1} - 1)} \|M_\phi(R_{k+n,k})\|$$

$$= \sqrt{2^{k+n+1}\ 2^{k+1}}\ \|M_\phi(R_{k+n,k})\| \quad .$$

Now $M_\phi(R_{j,k})$ has the form

$$M_\phi(R_{jk}) = \begin{bmatrix} a_{-2^{j+1}+1-2^k} & & a_{-2^{j+1}-2^{k+1}-2} \\ & \ddots & \\ a_{-2^j-2^k} & & a_{-2^j-2^{k+1}-1} \end{bmatrix}$$

FIGURE 2

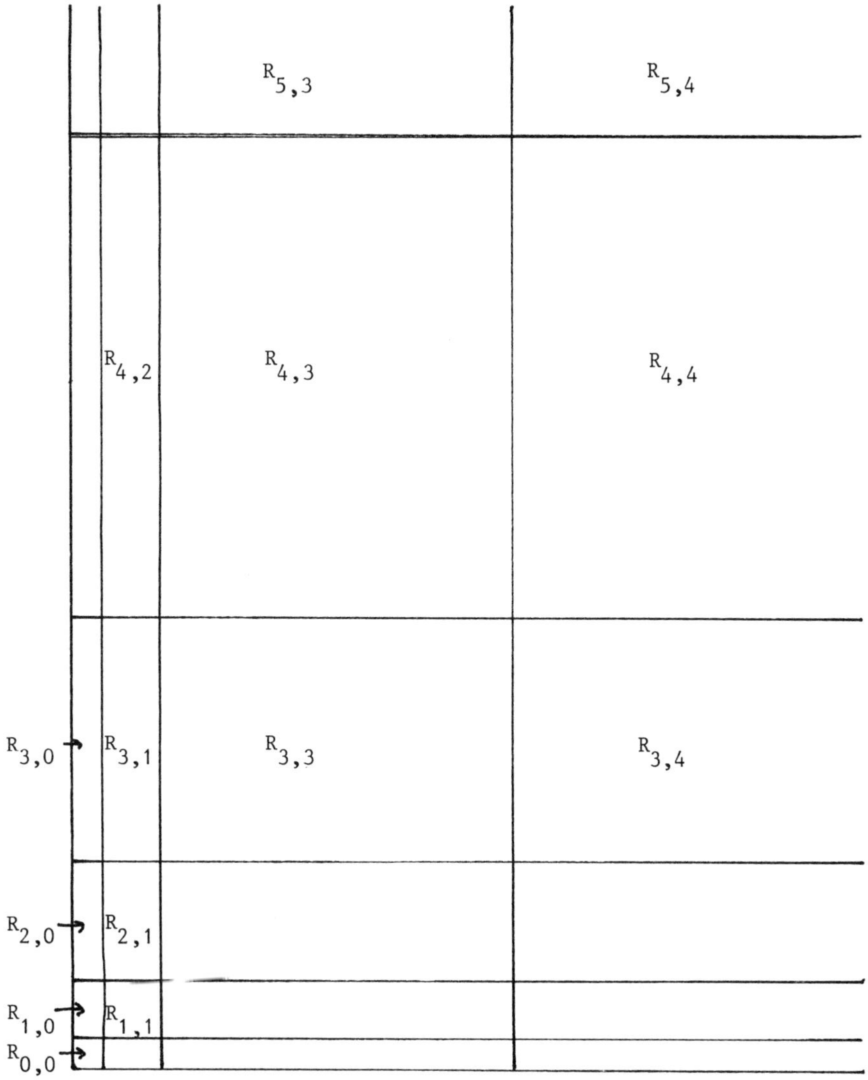

$$D_n = \bigcup_{i-j=n} R_{ij}$$

D_n is the "n th diagonal".

and is thus seen to be a submatrix of $T_{2^j+2^k}$. Applying the bound $\|T_\ell\| \leq \frac{K}{\ell}$ to $\|M_\phi(R_{k+n,n})\|$ gives

$$\|\tilde{M}_\phi(R_{k+n,k})\| \leq 2^{k+n+1}\, 2^{k+1} \frac{K}{2^{k+n} + 2^k}$$

$$\leq \frac{2K}{2^{n/2}} \quad .$$

Hence $\|\tilde{M}_\phi(D_n)\| \leq \frac{2K}{2^{n/2}}$.

If n is negative, a similar computation gives

$$\|\tilde{M}_\phi(D_n)\| \leq \frac{2K}{2^{|n|/2}} \quad .$$

Hence $\sum \|M_\phi(D_n)\| < \infty$. This shows that $\tilde{M}_{-+} = \sum \tilde{M}_\phi(D_n)$ is bounded.

Case 2: "M_{++} is bounded." The argument here is similar in spirit to the argument for Case 1, though with a new twist.

Partition $\mathbb{Z}_+ \times \mathbb{Z}_+$ into the set D_{-1} and the rectangles R^ℓ_{jk} $(j = 0,1,2,\ldots;\ \ell = 0,1)$ as shown in Figure 3. For each integer $n = 0,1,2,\ldots$ let $D^\ell_n = \cup_{k=0}^\infty R^\ell_{nk}$. The rectangles making up any diagonal D^ℓ_n are mutually orthogonal. Thus

$$\|M_\phi(D^\ell_n)\| = \sup_{k \geq 0} \|\tilde{M}_\phi(R^\ell_{n-k})\| \, . \qquad [*]$$

For each k, $M_\phi(R^\ell_{nk}) = M_\phi(R^\ell_{n0})$ and R^ℓ_{n0} is the rectangle $[0,2^j-1] \times [2^{j+1}-2+\ell 2^j,\ 2^{j+1}+2^j-3+\ell 2^j]$. Thus $M_\phi(R^\ell_{n0})$ is seen to be a submatrix of $T_{2^j-1+\ell 2^j}$, and the Factorization Lemma applied to * gives

FIGURE 3

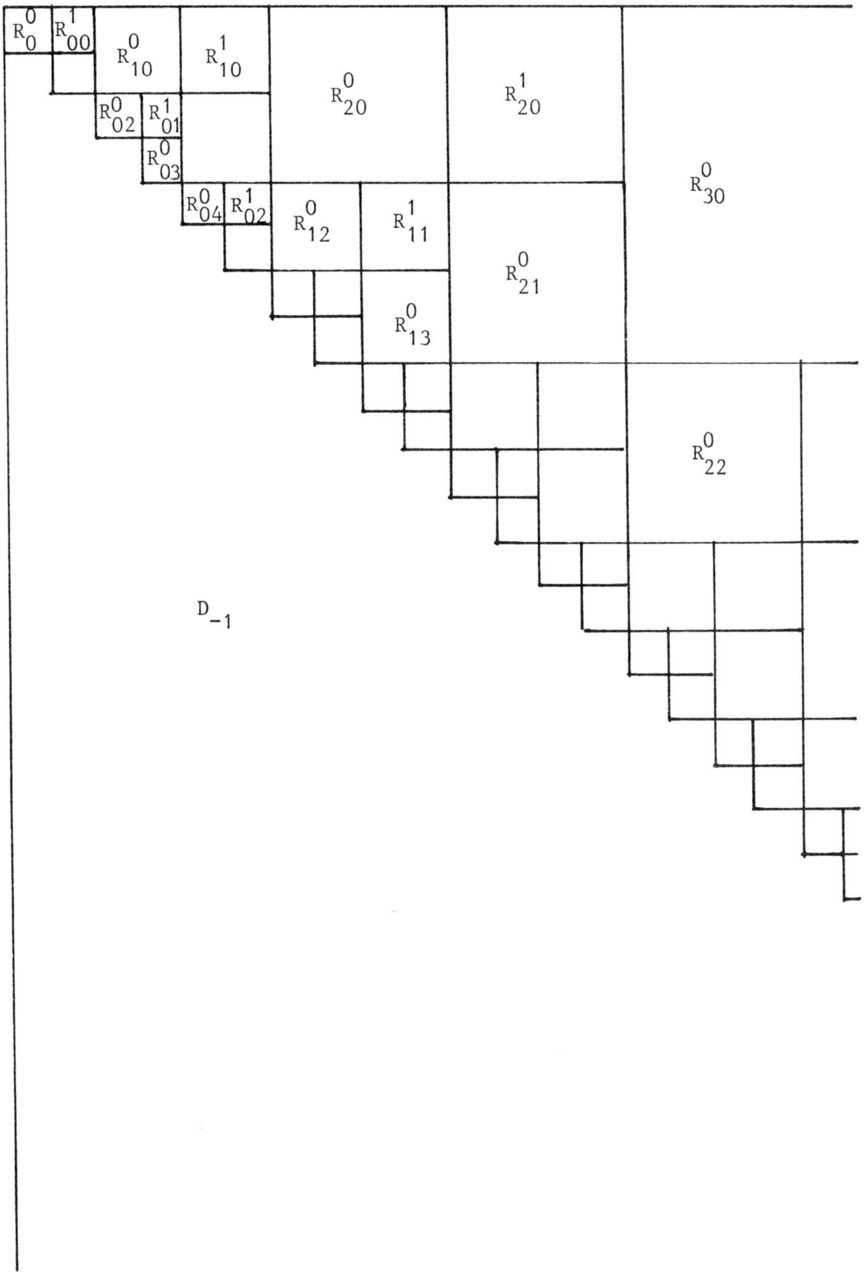
R^0_0
R^1_{00}
R^0_{10}
R^1_{10}
R^0_{20}
R^1_{20}
R^0_{02}
R^1_{01}
R^0_{03}
R^0_{30}
R^0_{04}
R^1_{02}
R^0_{12}
R^1_{11}
R^0_{21}
R^0_{13}
R^0_{22}
D_{-1}

$$\|\tilde{M}_\phi(D_n^\ell)\| \leq \frac{\beta(0)}{\beta(2^{j+1} + 2^j - 3 + \ell 2^j)} \|T_{2^j-1+\ell 2^j}\|$$

$$\leq \frac{\sqrt{2^{j+1} + 2^j - 3 + \ell 2^j + 1}}{2^j - 1 + \ell 2^j} K$$

$$\leq \frac{\sqrt{2^{j+2}}}{2^{j-1}} K$$

$$= \frac{2\sqrt{2}\,K}{2^{j/2}} \quad .$$

Thus $\sum_{\substack{n=0\\ \ell=0,1}}^{\infty} \|\tilde{M}_\phi(D_n^\ell)\| < \infty$ and it remains only to show that $\tilde{M}_\phi(D_{-1})$ is bounded.

By Proposition 4.4, $\bar{\phi}_{\bar{a}}(e^{i\theta}) = \sum_{n\leq 0} a_n e^{in\theta}$ is the Fourier series of a bounded function on T. By assumption $\phi \in L^\infty(T)$, so $\phi_a(e^{i\theta}) = \sum_{n>0} a_n e^{in\theta}$ is also the Fourier series of a bounded function on T. But $\tilde{M}_\phi(D_{-1}) = \tilde{M}_{\phi_a}(D_{-1})$, so it suffices to show that $\tilde{M}_{\phi_a}(D_{-1})$ is bounded. Proposition 2.1 shows that the matrix $(a_{jk})_{jk\in\mathbb{Z}_+}$ given by

$$a_{jk} = \begin{cases} 0 & j \leq k \\ \sqrt{\dfrac{k+1}{j+1}}\, a_{j-k} & j > k \end{cases}$$

is the matrix of the Bergman Toeplitz operator B_{ϕ_a} where here ϕ_a denotes the extension of $\phi_a : T \to \mathbb{C}$ to a bounded analytic function $\phi_a : D \to \mathbb{C}$. But $\tilde{M}_\phi(D_{-1})$ is the matrix $(c_{jk})_{jk\in\mathbb{Z}}$, where

$$c_{jk} = \begin{cases} \dfrac{\beta(j)}{\beta(k)}\, a_{j-k} & \text{if } j > k \geq 0 \;, \\ 0 & \text{otherwise.} \end{cases}$$

Finally, if $j > k \geqq 0$, then $\frac{\beta(j)}{\beta(k)} = \frac{\sqrt{k+1}}{\sqrt{j+1}}$. Thus $\tilde{M}_\phi(D_{-1})$ has the form of a bounded matrix embedded as a corner in a matrix which is zero elsewhere. So $\tilde{M}_\phi(D_{-1})$ is bounded.

Case 3: "M_{--} is bounded." The argument here is the same as for Case 2 and will be omitted.

Case 4: "M_{+-} is bounded." If E and G are the infinite matrices

$$E = \begin{bmatrix} \ddots & & & & \\ & 0 & & & \\ & & 0 & & \\ & & & \beta(0) & & \\ & & & & \beta(1) & \\ & & & & & \ddots \end{bmatrix},$$

and

$$F = \begin{bmatrix} \ddots & & & & & & \\ & \beta(-2)^{-1} & & & & & \\ & & \beta(-1)^{-1} & & & & \\ & & & \beta(0)^{-1} & & & \\ & & & & 0 & & \\ & & & & & 0 & \\ & & & & & & \ddots \end{bmatrix},$$

then E and F are bounded and

$$\tilde{M}_{+-} = E\, M_\phi\, F .$$

Therefore $\tilde{M}_{+-}$ is bounded. This completes the proof of the theorem. □

Let $\phi \sim \sum a_n e_n \in L^\infty(\beta)$. Lemma 5.3 says that ϕ may be identified with the bounded function on T, also denoted by ϕ, with Fourier expansion $\phi(e^{i\theta}) = \sum a_n e^{in\theta}$. This identification is a useful one and will be adopted here.

Write $\phi = \phi_a + \bar{\phi}_{\bar{a}}$, where $\phi_a(e^{i\theta}) = \sum_{n=0}^{\infty} a_n e^{in\theta}$ and $\phi_{\bar{a}}(e^{i\theta}) = \sum_{n=1}^{\infty} \bar{a}_{-n} e^{in\theta}$. ϕ_a and $\bar{\phi}_{\bar{a}}$ are the analytic and coanalytic parts of ϕ, respectively. The condition on the negative Fourier coefficients in Theorem 5.5 that $\sup n \|T_n\| < \infty$ was shown in Theorem 4.6 to be equivalent to $\phi_{\bar{a}}$ being in the space $B = B_1$. The next two propositions will show that this decomposition is a natural one in the sense that

$$\|\tilde{M}_\phi\| \cong \|\phi_a\|_\infty + \|\phi_{\bar{a}}\|_B .$$

<u>5.6 Proposition</u> If $\phi(e^{i\theta}) = \sum_{n=0}^{\infty} a_n e^{in\theta} \in B$ and if $\phi_N(e^{i\theta}) = \sum_{n=N}^{\infty} a_n e^{in\theta}$, then $\phi_N \in L^\infty(T)$ and

$$\|\phi_N\|_\infty \leq \frac{12 \log(N+1)}{N} \|\phi\|_B .$$

<u>Proof</u>. If T_{ϕ_N} denotes the Toeplitz operator of multiplication by ϕ_N on $H^2(T)$, then it suffices to show that $\|T_{\phi_N}\| \leq \frac{12 \log(N+1)}{N} \|\phi\|_B$. Let T be the matrix of T_{ϕ_N} with respect to the basis $\{e^{in\theta} : n = 0,1,2,\ldots\}$ of $H^2(T)$. T is constant along diagonals. Above the N th subdiagonal all entries are 0. Along the n th subdiagonal, for $n \geq N$, all entries are equal to a_n.

Partition that part of $\mathbb{Z}_+ \times \mathbb{Z}_+$ which corresponds to the non-zero entries of the matrix T into rectangles R^ℓ_{jk} $(j,k = 0,1,2,\ldots;\ \ell = 0,1)$ as indicated in Figure 4. If $j \geq 1$, then R_{jk} is a square of side length $2^{j-1}N$. $R^\ell_{0k} = R_{0k}$ is a square of side length N. If $n \geq 0$, define $D^\ell_n = \cup_{k\geq 0} R^\ell_{nk}$.

FIGURE 4

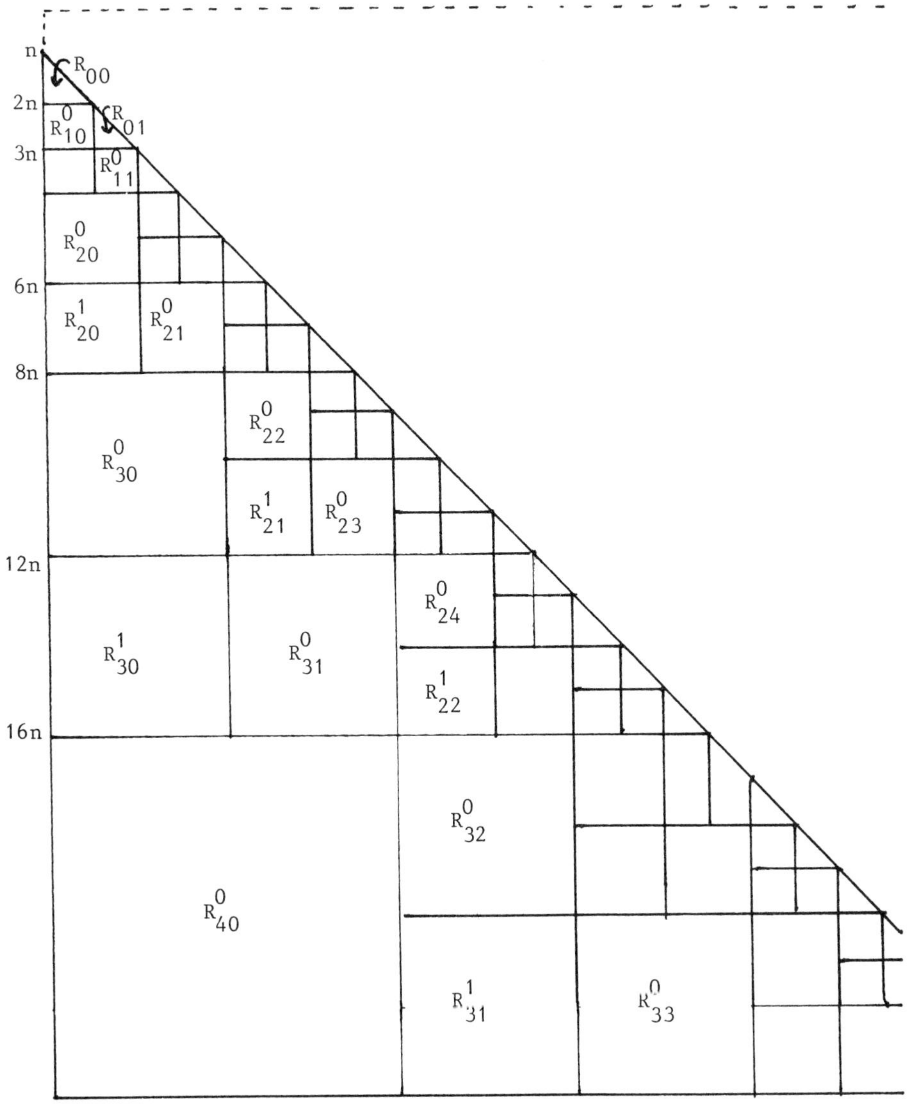
n
2n
3n
6n
8n
12n
16n
R_{00}
R_{01}
R^0_{10}
R^0_{11}
R^0_{20}
R^1_{20}
R^0_{21}
R^0_{22}
R^0_{30}
R^1_{21}
R^0_{23}
R^0_{24}
R^1_{30}
R^0_{31}
R^1_{22}
R^0_{32}
R^0_{40}
R^1_{31}
R^0_{33}

The rectangles which make up any given diagonal D_n^ℓ are mutually orthogonal. Moreover, because T is constant along diagonals $T(R_{nk_1}^\ell) = T(R_{nk_2}^\ell)$ for any choice of k_1, k_2. Therefore $\|T(D_n^\ell)\| = \|T(R_{n0}^\ell)\|$ for each n and ℓ.

If $n = 1,2,3,\ldots,$ then $T(R_{n0}^0)$ has the form

$$T(R_{n0}^0) = \begin{bmatrix} & & \cdot \\ a_{2^{n-1}N} & & a_{2^{n-2}N+1} \\ & \ddots & \\ a_{2^nN-1} & & a_{2^{n-1}N} \end{bmatrix} .$$

The transpose of $T(R_{n0}^0)$ (not the conjugate transpose) has the same norm as $T(R_{n0}^0)$ and is a submatrix of $T_{2^{n-2}N+1}$, where T is as in Definition 4.5. Since $\phi \in B_1$, there is a constant K such that $\|T_\ell\| \leq \frac{K}{\ell}$ for each ℓ. Therefore

$$\|T(D_n^0)\| = \|T(R_{n0}^0)\|$$

$$\leq \frac{K}{2^{n-2}N + 1}$$

Likewise, if $n \geqq 1$, then $\|T(D_n^1)\| \leq \frac{K}{2^{n-2}N + 1}$. Therefore

$$\sum_{\substack{n\geqq 1 \\ \ell=0,1}} \|T(D_n^\ell)\| \leq \frac{8K}{N} . \qquad [*]$$

All that remains is to estimate $\|T(R_{00}^0)\|$. Let $M = T(R_{00}^0)$. M is an $N\times N$ lower triangular matrix with the form

$$M = \begin{bmatrix} a_N & & & & 0 \\ a_{N+1} & a_N & & & \\ & \ddots & \ddots & & \\ & & \ddots & \ddots & \\ a_{2N} & & & a_{N+1} & a_N \end{bmatrix} .$$

Consider the partitioning of the lower triangular part of M into rectangles S_{jk} as indicated in Figure 5. The transpose of each matrix $M(S_{jk})$ is a submatrix of T_N. If n is the smallest integer so that $2^{n+1} \geqq N + 1$, then the indices j run from 0 to n. With the same logic as before, it follows that

$$\begin{aligned} \|M\| &\leqq \sum_{j=0}^{n} \|M(S_{j0})\| \\ &\leqq (n+1)\|T_N\| \\ &\leqq (n+1)\frac{K}{N} . \end{aligned}$$

But $n+1 < \log_2(2N) \leqq 4 \log(N)$, so

$$\|M\| \leqq \frac{4 \log(N+1)K}{N} .$$

Combining this with * gives the desired bound. □

5.7 Corollary There exist constants $c_1, c_2 > 0$ such that if $\phi \in L^\infty(\beta)$,

$$c_1\|\tilde{M}_\phi\| \leqq \|\phi_a\|_\infty + \|\phi_{\bar{a}}\|_B \leqq c_2\|\tilde{M}_\phi\| .$$

Proof. The proof of Theorem 5.5 shows that $2\,\|\tilde{M}_\phi\| \geqq \|\phi\|_\infty + \|\phi_{\bar{a}}\|_B$. Proposition 5.6 shows that $12\,\|\phi_{\bar{a}}\|_B \geqq \|\phi_{\bar{a}}\|_\infty$. Therefore

$$\begin{aligned}\|\phi_a\|_\infty &\leqq \|\phi\|_\infty + \|\phi_{\bar{a}}\|_\infty \\ &\leqq \|\phi\|_\infty + 12\,\|\phi_{\bar{a}}\|_B \quad .\end{aligned}$$

Hence $26\,\|\tilde{M}_\phi\| \geqq \|\phi_a\|_\infty + \|\phi_{\bar{a}}\|_B$. On the other hand, $\|\phi\|_\infty + \|\phi_{\bar{a}}\|_B$ was shown to be equivalent as a norm on $L^\infty(\beta)$ to the norm $\|\tilde{M}_\phi\|$. But

$$\begin{aligned}\|\phi\|_\infty &\leqq \|\phi_a\|_\infty + \|\phi_{\bar{a}}\|_\infty \\ &\leqq \|\phi_a\|_\infty + 12\,\|\phi_{\bar{a}}\|_B \quad .\end{aligned}$$

So for some constant $c > 0$,

$$\begin{aligned}\|M_\phi\| &\leqq c(\|\phi\|_\infty + \|\phi_{\bar{a}}\|_B) \\ &\leqq c(\|\phi_a\|_\infty + 13\,\|\phi_{\bar{a}}\|_B) \\ &\leqq 13c(\|\phi_a\|_\infty + \|\phi_{\bar{a}}\|_B) \quad .\end{aligned}$$

Just take $c_2 = 2$ and $c_1 = \frac{1}{13c}$. □

FIGURE 5

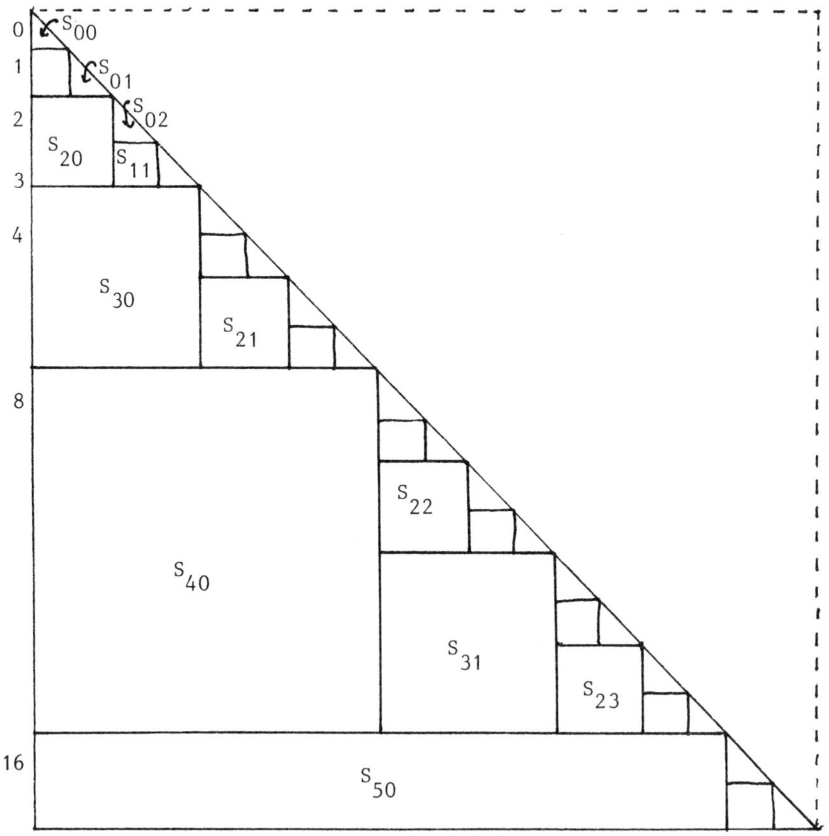

Decomposition of R_{00} for $N = 18$

§6. THE ALGEBRA $\{C_z\}'$

In the previous section, the commutant of C_z was identified to be equivalent to the algebra of functions $H^\infty \oplus \bar{B}$, where the multiplication is the pointwise product of functions on the circle T. In this section various questions about this algebra are posed, and it is shown that C_z has no square root. Finally, the maximal ideal space of $H^\infty \oplus B$ is identified with the corona of H^∞. First, the natural algebras generated by C_z and its inverse will be determined.

For a general, invertible bilateral shift S, it is known that the commutant of S coincides with the strongly closed algebra generated by S and S^{-1} (Shields [5], p. 62). In fact, if $L^\infty(\beta)$ denotes the algebra of Laurent series corresponding to $\{S\}'$, and if $\phi \in L^\infty(\beta)$, then ϕ is the strong limit of the Cesàro means of ϕ.

The strongly closed algebra generated by C_z and C_z^{-1} is therefore $\{C_z\}'$. Given the splitting of $\{C_z\}'$ into a direct sum of an "analytic" part and a "Besov" part, it follows that the strongly closed algebra generated by C_z and 1 is $H^\infty \oplus \{0\}$, and the strongly closed algebra generated by C_z^{-1} is $\{0\} \oplus \bar{B}$.

The article of Shields does not identify the norm closure of the algebras generated by C_z or C_z^{-1}. This will be done next.

Let $A(D)$ be the disc algebra consisting of all analytic functions on D which extend to continuous functions on the closed unit disc $\bar{D}$. It is well knwon that $A(D)$ is the uniform closure of the polynomials.

<u>6.1 Proposition</u> The norm closed algebra $A(C_z)$, generated by C_z and 1, is given by

$$A(C_z) = \{\tilde{M}_\phi : \phi \in A(D)\} .$$

Proof. $A(C_z)$ just corresponds to the norm closure of the polynomials in Z in $H^\infty \oplus \bar{B}$. This is the uniform closure of the polynomials in H^∞. □

For the next proposition, let $S_i = S_i'$ $(i = 1,2,3)$ be as defined at the end of Section Four.

6.2 Proposition The norm closed algebra $A(C_z^{-1})$ generated by 1 and C_z^{-1} has the descriptions:

$$\begin{aligned} A(C_z^{-1}) &= \{\tilde{M}_{\bar{\phi}} : \phi \in B_1 \text{ and } \lim_{n\to\infty} n\,\|T_n(\phi)\| = 0\} \\ &= \{\tilde{M}_{\bar{\phi}} : \phi \in B_1 \text{ and } \lim 2^n \|W_n * \phi\| = 0\} , \\ &= \{\tilde{M}_{\bar{\phi}} : \phi \in B \text{ and } \lim_{t\to 0^+} \frac{\|\Delta_t^2 \phi\|_\infty}{t} = 0\} . \end{aligned}$$

Proof. $A(C_z^{-1})$ corresponds to the norm closure of the polynomials in $\bar{z}$ in $H^\infty \oplus \bar{B}$. Now just apply Proposition 4.17. □

It is time to make good a promise of Section Three. The next proposition shows that C_z has no square root.

6.3 Proposition C_z has no square root.

Proof. Suppose to the contrary that $A^2 = C_z$. Since A commutes with C_z, write $A = \tilde{M}_\phi$ for some $\phi \in H^\infty \oplus \bar{B}$. Since C_z is invertible, so too is A, and A^{-1} commutes with C_z, so write $A^{-1} = \tilde{M}_\gamma$ for some

$\gamma \in H^\infty \oplus \bar{B}$. Now

$$\phi^2(e^{i\theta}) = e^{i\theta} \qquad \text{a.e.}$$

and

$$\phi(e^{i\theta})\gamma(e^{i\theta}) = 1 \qquad \text{a.e.}$$

Therefore $|\phi| = 1$ a.e. and $\gamma = \bar{\phi}$. But this says that both the analytic and coanalytic parts of ϕ are in B, hence are continuous. Therefore $\phi\colon T \to \mathbb{C}$ is a continuous function whose square is equal everywhere to $e^{i\theta}$. By index theory no such function exists. Thus C_z has no square root. □

Let $A = H^\infty \oplus B$ denote the algebra which corresponds to the commutant of C_z. If $n \geqq 0$, let e_n denote $e^{in\theta} \in H^\infty$. If $n < 0$, let e_n denote $e^{in\theta} \in \bar{B}$. It should be noted that for $n \geqq 0$, $\|e_n\| = 1$ and for $n < 0$, $\frac{|n|}{4} < \|e_n\| < |n|$.

Finally, the maximal ideal space of A will be determined. Recall that for an algebra B, $\hat{B}$ denotes the collection of multiplicative linear functionals on B. Of course there is an identification of the maximal ideal space with $\hat{B}$. The identification takes each linear functional to the ideal which is its kernel. Here it will be shown that the maximal ideal space of A is identical with the corona of H^∞. The corona of H^∞ is the collection of maximal ideals m in $\hat{H}^\infty$ for which $|m(e_1)| = 1$.

The first step along the way is a description of the spectrum of elements in B. B can be viewed as an algebra of bounded analytic functions on the disc whose extensions to the closed disc are continuous. It is therefore obvious that if $\varphi \in B$, then $\sigma_B(\varphi) \supset \varphi(\text{cl } D)$. In fact equality holds here.

<u>6.4 Proposition</u> If $\varphi \in B$, then $\sigma_B(\varphi) = \varphi(\text{cl } D)$.

Proof. It suffices to show that $(\varphi-\lambda)^{-1}$ is in B for each complex number λ a positive distance, say ε, or more away from $\varphi(D)$. For such λ it suffices to show that $\|\Delta_\eta^2(\varphi-\lambda)^{-1}\|_\infty$ converges with appropriate rapidity to zero as η tends to zero. A straightforward computation reveals that

$$(\Delta^2(\varphi-\lambda)^{-1})\ (\theta) = C(\theta,\eta,\lambda)^{-1}\ [\lambda(\Delta_\eta^2\varphi)(\theta) + R(\theta,\eta)]$$

where $C(\theta,\eta,\lambda) = \prod_{\delta=0,\pm\eta} (\varphi(\theta+\delta)-\lambda)$, and where R is independent of λ. For λ at least ε away from $\varphi(\mathrm{cl}\ D)$, $C(\theta,\eta,\lambda)^{-1}$ is uniformly bounded. Therefore the behavior of the sum $\lambda(\Delta_\eta^2\varphi)(\theta) + R(\theta,\eta)$ is of interest. Since φ is in B, only the second term is of interest here. But as R is independent of λ and $(\varphi-\lambda)^{-1}$ is in B for some sufficently large value of λ, it follows that $R(\theta,\eta)$ converges to zero fast enough to make $(\varphi-\lambda)^{-1}$ be in B for all λ away from $\varphi(\mathrm{cl}\ D)$. □

6.5 Corollary $\hat{B} = \mathrm{cl}\ D$.

Proof. If $\lambda \in \mathrm{cl}\ D$ then $m_\lambda: B \to \mathbb{C}$ defined by $m_\lambda(\varphi) = \varphi(\lambda)$ is certainly a multiplicative linear functional on B. Next suppose $m: B \to \mathbb{C}$ is a given multiplicative linear functional and set $\lambda = m(e_1)$. It must be shown that $m(\varphi) = \varphi(\lambda)$ for each φ in B. So fix φ in B. The previous proposition shows that $|m(\varphi)| \leqq \|\varphi\|_\infty$. Polynomials are unformly dense in B, even if they are not dense in the norm of B, so choose polynomials P_n so that $\|P_n-\varphi\|_\infty \to 0$. Clearly $m(P_n) = P_n(\lambda)$ and therefore

$$\begin{aligned}|m(\varphi)-\varphi(\lambda)| &= \lim_n |m(\varphi)-P_n(\lambda)| \\ &= \lim_n |m(\varphi)-m(P_n)| \\ &= \lim |m(\varphi-P_n)| \\ &\leqq \lim\sup \|\varphi-P_n\|_\infty \\ &= 0\end{aligned}$$

□

<u>6.6 Proposition</u> There is a natural injection

$$\text{Corona}\ (H^\infty) \to \hat{A}\ .$$

<u>Proof</u>. If $m \in \text{Corona}(H^\infty)$, let $\lambda = m(e_1)$. Since m is in the Corona, $|\lambda| = 1$. Define a function $m'\colon A \to \mathbb{C}$ by setting

$$m'(f \oplus \bar{g}) = m(f) + \overline{g(\lambda)}$$

for each $f \in H^\infty$, $g \in B$. Since each element of B can be identified with a continuous function on T, the map is well defined. m' is easily seen to be bounded, since if $g \in B$,

$$|m'(\bar{g})| = |g(\lambda)| \leqq \|g\|_\infty \leqq \|g\|_B \quad .$$

The restrictions of m' to H^∞ and B are also seen to be multiplicative. It remains only to show that $m'(f\bar{g}) = m'(f)m'(g)$ for each $f \in H^\infty$ and $g \in B_1$.

Assume $f \in H^\infty$ has the expansion $\sum_{n \geqq 0} a_n e_n$. For $N \in \mathbb{Z}_+$ let $f_N = \sum_{n \geqq N} a_n e_n$. Since f_N and the product $f_N e_{-N}$ are in H^∞, it follows that

$$\begin{aligned} m'(f_N e_{-N}) &= m'(f_N)(m'(e_N))^{-1} \\ &= m'(f_N)\lambda^{-N} \quad . \end{aligned}$$

By direct computation $(f-f_N$ is a polynomial), it follows that

$$m'((f-f_N)e_{-N}) = \lambda^{-N}\ m(f-f_N) \quad .$$

Combining these results gives $m'(fe_{-N}) = m'(f)m'(e_N)$. Hence, for any polynomial P, and $f \in H^\infty$,

$$m'(f\bar{P}) = m'(f)m'(\bar{P}) \quad .$$

Now suppose $f \in H^{\infty}$ and $g \in B$. Viewed as a function on the circle, g is continuous. If $\varepsilon > 0$, choose a polynomial P so that $\|g-P\|_{L^{\infty}(T)} < \varepsilon$. Then

$$\begin{aligned} |m'(f\bar{g}) - m'(f)m'(\bar{g})| &= |m'(f\bar{P}) + m'(f\overline{g-P}) - m'(f)m'(\bar{P}) - m'(f)m'(\overline{g-P})| \\ &\leqq |m'(f\overline{g-P})| + |m'(f)m'(\overline{g-P})| \\ &\leqq 2\,\|f\|_{\infty}\,\|g-P\|_{\infty} \\ &\leqq 2\,\|f\|_{\infty}\,\varepsilon \quad . \end{aligned}$$

Clearly $m'(f\bar{g}) = m'(f)m'(\bar{g})$. This shows that m' is in fact multiplicative. □

<u>6.7 Proposition</u> There is a natural projection

$$\hat{A} \to \text{Corona}\,(H^{\infty})$$

given by the restriction map.

<u>Proof</u>. If $m \in A$, then it must be shown that $|m(e_1)| = 1$. Let $\lambda = m(e_1)$. Clearly $|\lambda| \leqq 1$. By the boundedness of m,

$$\sup_{n \in \mathbb{Z}} \frac{|m(e_n)|}{\|e_n\|} \leqq 1 \quad .$$

If $n > 0$, then $m(e_{-n}) = \lambda^{-n}$ and $\|e_{-n}\| < n$. Therefore

$$\sup_{n>1} \frac{|\lambda|^{-n}}{n} \leqq 1 \quad .$$

Hence $|\lambda| \geqq 1$. This shows $|\lambda| = 1$. □

Proposition 6.7 says that the restriction map $m \to m|H^\infty$ goes from $\hat{A}$ onto Corona(H^∞). Proposition 6.6 says that the map is surjective. Corollary 6.5 says that m can be recovered from $m|H^\infty$, as the only multiplicative linear functionals on B are the point evaluations. In other words the map is also injective.

6.8 Theorem The restriction map $\hat{A} \to$ Corona(H^∞) is an isometric surjection. □

§7. FINAL REMARKS

In this final section some comments will be made as to the results and methods of this work.

With no additional work the methods of Section Five can be used to obtain an identification of $L^\infty(\beta)$ for certain other sequences $\beta: \mathbb{Z} \to \mathbb{C}$. If $\alpha > 0$, define $\beta_\alpha : \mathbb{Z} \to \mathbb{C}$ by

$$\beta_\alpha(n) = \begin{cases} (n+1)^{-\alpha/2} & \text{if } n \geq 0 , \\ (|n|+1)^{\alpha/2} & \text{if } n < 0 . \end{cases}$$

This sequence β_α corresponds to the weight sequence $w_\alpha : \mathbb{Z} \to \mathbb{C}$ given by

$$w_\alpha(n) = \begin{cases} (\frac{n+1}{n+2})^{\alpha/2} & \text{if } n \geq 0 , \\ (\frac{|n|}{|n|+1})^{\alpha/2} & \text{if } n < 0 . \end{cases}$$

Recall that $L^\infty(\beta_\alpha)$ corresponds to the commutant of the bilateral shift with weighted sequence w_α. The Bergman bilateral shift corresponds to the case $\alpha = 1$. The methods of Section Five now can be used to give the identification

$$L^\infty(\beta_\alpha) \cong H^\infty \oplus B_\alpha .$$

The work of Peller [7] has relevance here. Recall that in the proof of Theorem 5.5 the matrix of $\tilde{M}_\phi$ was broken into four submatrices, one of which was labelled $\tilde{M}_{-+}$. Peller has determined necessary and sufficient conditions on a formal Laurent series ϕ, which describe when $\tilde{M}_{-+}$ is the matrix of an operator in a Schatten-p class. The condition is that $\phi_{\bar{a}}$ be in a certain Besov space. His methods apply also to the more general sequences β_α. Though Peller's approach is more powerful than the one outlined here, it does not seem capable of handling certain "β-sequences" with less structure than those given here. For certain such sequences the methods outlined here give much information as to the nature of $L^\infty(\beta)$.

It is also possible to show that $L^\infty(\beta)$ is stable in some sense with respect to perturbations of the sequences β. Thus if $\beta' : \mathbb{Z} \to \mathbb{C}$ is another sequence such that

$$\sup\{\left|\frac{\beta(n)}{\beta'(n)}\right| , \left|\frac{\beta'(n)}{\beta(n)}\right| : n \in \mathbb{Z}\} < \infty \quad ,$$

then as sets of formal Laurent series $L^\infty(\beta) = L^\infty(\beta')$.

The structure of the lattice of invariant subspaces of the Bergman shift (and hence by heredity the bilateral Bergman shift) is hopelessly complex. One could hope however to use the description of the commutant of C_z to determine the structure of its lattice of hyperinvariant subspaces.

Finally it is hoped that a more detailed study of the bilateral Bergman shift could lead to information about the spectrum of Bergman-Toeplitz operators.

REFERENCES

[1] S. Axler, "Multiplication Operators on Bergman Spaces". J. für die reine u. angewandte Mathematik 336 (1982), 26-44

[2] J.B. Conway, "Subnormal Operators", Pitman, Advanced Publications Program, Boston, 1981.

[3] R. Douglas, "Banach Algebra Techniques in Operator Theory", Academic Press, New York, 1972.

[4] G. McDonald and C. Sunberg, "Toeplitz Operators on the Disc", Indiana University Math. J. 28 (1979), 595-611.

[5] C. Pearcy (ed.), "Topics in Operator Theory", American Math. Society Providence, RI, 1974.

[6] J. Peetre, "New Thoughts on Besov Spaces", Duke University Mathematics Series, Durham, NC (1976).

[7] V.V. Peller, "Vectorial Hankel Operators, Commutators, and Related Operators of the Schatten-von Neuman Class γ_p", LOMI Preprint, Leningrad, 1981.

[8] P. Brenner, V. Thomée, L.B. Wahlbin, "Besov Spaces and Applications to Difference Methods for Initial Value Problems", Springer-Verlag.

Gregory Adams, Guest at Technische Hochschule Darmstadt (West Germany)

General instructions to authors for
PREPARING REPRODUCTION COPY FOR MEMOIRS

For more detailed instructions send for AMS booklet, "A Guide for Authors of Memoirs." Write to Editorial Offices, American Mathematical Society, P. O. Box 6248, Providence, R. I. 02940.

MEMOIRS are printed by photo-offset from camera copy fully prepared by the author. This means that, except for a reduction in size of 20 to 30%, the finished book will look exactly like the copy submitted. Thus the author will want to use a good quality typewriter with a new, medium-inked black ribbon, and submit clean copy on the appropriate model paper.

Model Paper, provided at no cost by the AMS, is paper marked with blue lines that confine the copy to the appropriate size. Author should specify, when ordering, whether typewriter to be used has PICA-size (10 characters to the inch) or ELITE-size type (12 characters to the inch).

Line Spacing – For best appearance, and economy, a typewriter equipped with a half-space ratchet – 12 notches to the inch – should be used. (This may be purchased and attached at small cost.) Three notches make the desired spacing, which is equivalent to 1-1/2 ordinary single spaces. Where copy has a great many subscripts and superscripts, however, double spacing should be used.

Special Characters may be filled in carefully freehand, using dense black ink, or INSTANT ("rub-on") LETTERING may be used. AMS has a sheet of several hundred most-used symbols and letters which may be purchased for $5.

Diagrams may be drawn in black ink either directly on the model sheet, or on a separate sheet and pasted with rubber cement into spaces left for them in the text. Ballpoint pen is *not* acceptable.

Page Headings (Running Heads) should be centered, in CAPITAL LETTERS (preferably), at the top of the page – just above the blue line and touching it.

LEFT-hand, EVEN-numbered pages should be headed with the AUTHOR'S NAME;

RIGHT-hand, ODD-numbered pages should be headed with the TITLE of the paper (in shortened form if necessary).

Exceptions: PAGE 1 and any other page that carries a display title require NO RUNNING HEADS.

Page Numbers should be at the top of the page, on the same line with the running heads.

LEFT-hand, EVEN numbers – flush with left margin;

RIGHT-hand, ODD numbers – flush with right margin.

Exceptions: PAGE 1 and any other page that carries a display title should have page number, centered below the text, on blue line provided.

FRONT MATTER PAGES should be numbered with Roman numerals (lower case), positioned below text in same manner as described above.

MEMOIRS FORMAT

It is suggested that the material be arranged in pages as indicated below.
Note: Starred items (*) are requirements of publication.

Front Matter (first pages in book, preceding main body of text).

Page i – *Title, *Author's name.

Page iii – Table of contents.

Page iv – *Abstract (at least 1 sentence and at most 300 words).

**1980 Mathematics Subject Classification (1985 Revision).* This classification represents the primary and secondary subjects of the paper, and the scheme can be found in Annual Subject Indexes of MATHEMATICAL REVIEWS beginning in 1984.

Key words and phrases, if desired. (A list which covers the content of the paper adequately enough to be useful for an information retrieval system.)

Page v, etc. – Preface, introduction, or any other matter not belonging in body of text.

Page 1 – Chapter Title (dropped 1 inch from top line, and centered).

Beginning of Text.

Footnotes: *Received by the editor date.

Support information – grants, credits, etc.

Last Page (at bottom) – Author's affiliation.

ABCDEFGHIJ – 89876